SpringerBriefs in Physics

SpringerBriefs in Physics are a series of slim high-quality publications encompassing the entire spectrum of physics. Manuscripts for SpringerBriefs in Physics will be evaluated by Springer and by members of the Editorial Board. Proposals and other communication should be sent to your Publishing Editors at Springer.

Featuring compact volumes of 50 to 125 pages (approximately 20,000–45,000 words), Briefs are shorter than a conventional book but longer than a journal article. Thus, Briefs serve as timely, concise tools for students, researchers, and professionals.

Typical texts for publication might include:

- A snapshot review of the current state of a hot or emerging field
- A concise introduction to core concepts that students must understand in order to make independent contributions
- An extended research report giving more details and discussion than is possible in a conventional journal article
- A manual describing underlying principles and best practices for an experimental technique
- An essay exploring new ideas within physics, related philosophical issues, or broader topics such as science and society

Briefs allow authors to present their ideas and readers to absorb them with minimal time investment. Briefs will be published as part of Springer's eBook collection, with millions of users worldwide. In addition, they will be available, just like other books, for individual print and electronic purchase. Briefs are characterized by fast, global electronic dissemination, straightforward publishing agreements, easy-to-use manuscript preparation and formatting guidelines, and expedited production schedules. We aim for publication 8–12 weeks after acceptance.

Nima Moshayedi

Quantum Field Theory and Functional Integrals

An Introduction to Feynman Path Integrals and the Foundations of Axiomatic Field Theory

 Springer

Nima Moshayedi 🆔
Institute for Mathematics
University of Zurich
Zurich, Switzerland

ISSN 2191-5423   ISSN 2191-5431 (electronic)
SpringerBriefs in Physics
ISBN 978-981-99-3529-1   ISBN 978-981-99-3530-7 (eBook)
https://doi.org/10.1007/978-981-99-3530-7

This Springer imprint is published by the registered company Springer Nature Singapore Pte Ltd.
The registered company address is: 152 Beach Road, #21-01/04 Gateway East, Singapore 189721, Singapore

To My Daughter Luna

Preface

This book started as lecture notes for a course on quantum field theory and functional integrals at the University of Zurich. The notes have developed through the years and several things have been changed, revised and rewritten.

We describe the functional integral (Feynman path integral) approach to quantum mechanics and quantum field theory, where the main focus lies in Euclidean field theory. The notion of the Gaussian measure and the construction of the Wiener measure are covered. Moreover, we recall the notion of classical mechanics and the Schrödinger picture of quantum mechanics, where we show the equivalence to the path integral formalism, by deriving the quantum mechanical propagator from it. Additionally, we give an introduction to elements of constructive quantum field theory.

We start with a brief recap of classical mechanics for the inexperienced reader, where we cover the basic notions from Newtonian mechanics, Hamiltonian mechanics and Lagrangian mechanics. Then we move on to the Schrödinger picture of quantum mechanics, where we cover the main postulates and recall some basic notions of functional analysis. Then we continue with the Weyl quantization picture and introduce some aspects of Fourier analysis. Moreover, we use the Fourier transform in order to derive a solution to the Schrödinger equation.

Then we move on to Feynman's path integral quantization approach, where we consider the free propagator and introduce the notion of Gaussian measures and some of their properties. Furthermore, we discuss the notion of Feynman diagrams and how Wick's theorem is used in this case.

The last chapter considers aspects of constructive quantum field theory, where we consider some of the important constructions such as the Osterwalder–Schrader or the Garding–Wightman axioms.

This book is meant to be an introductory course on the analytical aspects of Euclidean quantum field theory. As already mentioned, it is based on a lecture and thus not every single notion is defined in full detail. It is also required that the reader is familiar with the basic aspects of differential geometry and functional analysis, even though the most basic aspects of functional analysis are recalled. This book has no pretence of originality.

Zurich, Switzerland   Nima Moshayedi

Acknowledgements I want to thank my advisor A. S. Cattaneo for helpful discussions, comments on the notes and his constant support. I also want to thank S. Kandel for sharing some of his handwritten notes with me. Moreover, I want to thank M. Nakamura for all the good correspondence and his help to publish these notes as a book in this series. Furthermore, I want to thank the referee for the comments which helped to improve this book. I also want to thank my wife Manuela for supporting me, especially in difficult situations, and for cheering me up whenever I did not feel very motivated.

Contents

Chapter 1
Introduction

What is This Book About?

In theoretical and mathematical physics, we are often interested in the transition from the *classical* to the *quantum* picture of mechanics. In the classical setting, we usually consider space-time, formed by some Euclidean space $\mathbb{R}^n$ of dimension n, and consider a state of a mass particle moving in space-time as an element of the *phase space*, i.e., an element of Euclidean space $\mathbb{R}^{2n}$ of dimension $2n$, hence twice the dimension we started with. The first n entries are the position coordinates $x_1, \ldots, x_n$ in space and the other n entries represent the momentum coordinates $p_1, \ldots, p_n$ associated with each position coordinate x_i for all $i = 1, \ldots, n$. *Observables* are given by smooth functions f on the phase space, i.e., $f \in C^\infty(\mathbb{R}^{2n})$. By time-evolution, each position and momentum coordinate of the considered mass particle changes and thus everything depends on the time t. So, in fact, we have $(x_1(t), \ldots, x_n(t), p_1(t), \ldots, p_n(t))$. The dynamics of observables is captured by the equation

$$\frac{df}{dt} = \{f, H\},$$

where $H \in C^\infty(\mathbb{R}^{2n})$ is some special function, called the *Hamiltonian function*, that describes the total energy. The bracket $\{\cdot, \cdot\}$ is a special $\mathbb{R}$-bilinear map $C^\infty(\mathbb{R}^{2n}) \times C^\infty(\mathbb{R}^{2n}) \to C^\infty(\mathbb{R}^{2n})$, called a *Poisson bracket*. It is also of crucial importance for the dynamical information of the system. When passing to the quantum picture, it is important to understand the corresponding objects in this setting. There the observables are given by self-adjoint operators on some Hilbert space $\mathcal{H}$ and a quantum state is described by an element $\psi \in \mathcal{H}$. The dynamics are encoded in the *Schrödinger equation*

$$i\hbar \frac{\partial \psi}{\partial t} = \widehat{H}\psi,$$

© The Author(s), under exclusive license to Springer Nature Singapore Pte Ltd. 2023
N. Moshayedi, *Quantum Field Theory and Functional Integrals*,
SpringerBriefs in Physics, https://doi.org/10.1007/978-981-99-3530-7_1

where $\widehat{H} \in \text{End}(\mathcal{H})$ denotes the *Hamiltonian operator* of the system. Here we have denoted by $\text{End}(\mathcal{H}) := \{A : \mathcal{H} \to \mathcal{H} \mid A \text{ linear}\}$ the endomorphisms on $\mathcal{H}$. The dynamical equation for a quantum observable $A \in \text{End}(\mathcal{H})$ is given by the *Heisenberg equation*

$$\frac{dA}{dt} = \frac{i}{\hbar}[\widehat{H}, A],$$

where $[\cdot, \cdot]$ denotes the Lie bracket (commutator) on $\text{End}(\mathcal{H})$. The question that is considered in this book is about the mathematical transition of the classical picture to the quantum picture. Generally, such a procedure is called a *quantization*. We will look at particular ways in which to quantize certain mathematical structures in order to make the theory coherent.

A More Sophisticated Picture

In the *Lagrangian picture* of classical mechanics we consider an action functional[1] of the form

$$S[x] = \int_{t_0}^{t_1} L(x(t), \dot{x}(t)) dt,$$

where $L(x, \dot{x}) = \frac{1}{2}m\|\dot{x}\|^2 - V(x)$ is called the *Lagrangian function* (sometimes just *Lagrangian*) of a path $x : [t_0, t_1] \to \mathbb{R}^n$ with some function $V \in C^\infty(\mathbb{R}^n)$ depending on the path x, called the *potential energy*. The expression $T := \frac{1}{2}m\|\dot{x}\|^2$ is usually called the *kinetic energy*, where $m \in \mathbb{R}_{>0}$ is a positive scalar (the *mass* of the considered particle along the path x) and $\|\cdot\|$ denotes the Euclidean norm on $\mathbb{R}^n$. We denote by $P(\mathbb{R}^n, x_0, x_1)$ the space of all paths $x : [t_0, t_1] \to \mathbb{R}^n$ with $x(t_0) = x_0$ and $x(t_1) = x_1$. The *principle of least action* says that the classical trajectory (path) that the considered particle will choose is the one that minimizes the action functional. By considering the methods of *variational calculus*, one can show that the solutions of the equation $\delta S = 0$ for fixed endpoints (i.e., the critical points of S) give us the classical trajectory of the particle with mass $m \in \mathbb{R}_{>0}$. The equations following from $\delta S = 0$ are called the *Euler–Lagrange (EL) equations* and they are identical to the equations of motion obtained from the methods of *Newtonian mechanics*. Newton's equations of motion appear from the law $F = ma(t) = m\ddot{x}(t)$, where $a(t)$ denotes the *acceleration* defined as the second time-derivative of the path $x(t)$ (read as "force equals mass times acceleration"). To see this, we recall that *momentum* in physics is given by $p(t) = mv(t)$, where $v(t) = \dot{x}(t)$ denotes the velocity of the particle with mass m. Then, by the fact that the velocity is given by the first time-derivative of the path, i.e., $v = \dot{x}$, one considers the coordinates $\dot{x} = \frac{p}{m}$ and $\dot{p} = -\nabla V$, where $\nabla = (\partial_1, \ldots, \partial_n)^T$ denotes[2] the *gradient* operator on $\mathbb{R}^n$. The

---

[1] In the physics literature, it is common to denote the time-derivatives by "dots", i.e., $\dot{x}(t) := \frac{d}{dt}x(t)$.
[2] For local coordinates (x^i) we write $\partial_i := \frac{\partial}{\partial x^i}$.

Hamiltonian approach to classical mechanics considers the space with the coordinates (x, p) to be the classical *phase space* (classical space of *states*) given by the *cotangent bundle* $T^*\mathbb{R}^n \cong \mathbb{R}^{2n} \ni (x, p)$ endowed with a *symplectic form*[3] given by the closed and non-degenerate differential 2-form

$$\omega = \sum_{i=1}^{n} \mathrm{d}x^i \mathrm{d}p_i.$$

As already mentioned before, for a given classical system we consider the Hamiltonian function $H \in C^\infty(\mathbb{R}^n)$ which is given by the total energy $H(x, p) = \frac{\|p\|^2}{2m} + V(x)$, where $V \in C^\infty(\mathbb{R}^n)$ is again a potential energy function. In the physics literature, the first term of H is called the *kinetic energy*. This function is said to be *Hamiltonian* if there is a vector field X_H such that

$$\iota_{X_H}\omega = -\mathrm{d}H,$$

where ι denotes the *contraction* map (also called *interior derivative*). The vector field X_H is called the *Hamiltonian vector field* of H. In the case at hand, since ω is symplectic and hence non-degenerate, every function is Hamiltonian and its Hamiltonian vector field is uniquely determined. For H being the total energy function and ω the canonical symplectic form on the cotangent space, we get the following Hamiltonian vector field: A vector field on $T^*\mathbb{R}^n$ has the general form[4] $X = X^i \partial_{x^i} + X_i \partial_{p_i}$. Thus, applying the equation for the Hamiltonian vector field of H, we get $-\mathrm{d}H = X_i \mathrm{d}x^i + X^i \mathrm{d}p_i = \iota_X \omega$. Now since $\mathrm{d}H = \partial_i V \mathrm{d}x^i + \frac{p_i}{m}$, we get the coefficients of the vector field to be $X_i = -\partial_i V$ and $X^i = \frac{p_i}{m}$. Hence, we get the Hamiltonian vector field

$$X_H = \frac{p_i}{m}\partial_{x^i} - \partial_i V \partial_{p_i}.$$

Naturally, X_H induces a *Hamiltonian flow* which is a map $T^*\mathbb{R}^n \to T^*\mathbb{R}^n$, where the flow equations are given by

$$\dot{x}^i = \frac{p_i}{m},$$
$$\dot{p}_i = -\partial_i V.$$

An approach of quantization of the above is to associate with $T^*\mathbb{R}^n$ the Hilbert space of *square-integrable functions* $\mathcal{H}_0 = L^2(\mathbb{R}^n) := \{f \in C^\infty(\mathbb{R}^n) \mid \int_{\mathbb{R}^n} |f|^2 \mathrm{d}^n x < \infty\}$ on $\mathbb{R}^n$. The Hamiltonian flow can then be replaced by a linear map

---

[3] We will not always write the exterior product $\wedge$ between differential forms but secretly always mean that the exterior product is there between them, i.e., for two differential forms α, β, we will write $\alpha\beta$ when we actually mean $\alpha \wedge \beta$.

[4] Here we have written $\partial_x := \frac{\partial}{\partial x}$ and we impose the Einstein summation convention. This means that whenever there are repeating indices, we will automatically sum over those indices.

$$e^{\frac{i}{\hbar}\hat{H}} : L^2(\mathbb{R}^n) \longrightarrow L^2(\mathbb{R}^n),$$

where $\hat{H} := -\frac{\hbar^2}{2m}\Delta + V$ denotes the *Hamiltonian operator*, which appears as the canonical *operator quantization* of the classical Hamiltonian function H, where $\Delta = \sum_{1 \leq j \leq n} \partial_j^2$ denotes the *Laplacian* on $\mathbb{R}^n$. Note that the space of states is now given by a Hilbert space $\mathcal{H}_0$ and the observables as operators on $\mathcal{H}_0$. One can show that the action of this operator can be expressed as an integral of the form

$$\left(e^{\frac{i}{\hbar}\hat{H}}\psi\right)(x_0) = \int_{\mathbb{R}^n} K(x_0, x_1)\psi(x_1)dx_1,$$

for $\psi \in \mathcal{H}_0$, where K denotes the *integral kernel* for the operator $e^{\frac{i}{\hbar}\hat{H}}$. Feynman showed in Feynman (1942) that this kernel (quantum mechanical *propagator*) can be seen as a *path integral*, i.e., an object of the form

$$K(x_0, x_1) = \int_{P(\mathbb{R}^n, x_0, x_1)} e^{\frac{i}{\hbar}S[x]}\mathscr{D}[x],$$

where S denotes the action of the classical system and $\mathscr{D}$ a formal (non-existing) measure on the path space. Since $\mathscr{D}$ is supposed to be the "Lebesgue measure" on an infinite-dimensional space, it is mathematically ill-defined. However, one can still make sense of such an integral in several ways; one of them is by considering its *perturbative expansion* into a formal power series in $\hbar$ with coefficients given by weights assigned to certain graphs called *Feynman diagrams*. It actually turns out that this procedure is mathematically well-defined and gives surprisingly accurate predictions in modern theoretical physics. In fact, this procedure yields formal power series which one interprets as asymptotic series after which appropriate techniques (like the *Borel summation*) are used to get numerical predictions. An important phenomenon in theoretical physics related to this is called *resurgence*. These types of integrals are usually called *functional integrals* or *Feynman path integrals*.[5] This book will mainly emphasize on the quantization technique through these integrals. In particular, we will look at the quantum mechanical case with respect to these objects and consider some mathematical elements for the more advanced methods to describe *quantum field theory*. We will actually consider the axiomatic approach and describe some of the relevant constructions. This book is mainly based on Arnold (1978), Bogachev (1998), Eldredge (2016), Glaser (1974), Glimm and Jaffe (1987), Guichardet (1972), Hall (2013), Anson (1987), Johnson and Lapidus (2000), Kuo (1975), Mörters and Peres (2010), Osterwalder and Schrader (1975), Reed and Simon (1981), Simon (1974, 1979), Takhtajan (2008), Wightman and Garding (1964).

---

[5] To be precise, the name *path integral* refers to the 1-dimensional case (quantum mechanics) and *functional integral* to the higher-dimensional case (quantum field theory).

Chapter 2
A Brief Recap of Classical Mechanics

We want to start by recalling some of the most important notions of classical mechanics. We also want to mention the excellent reference (Arnold 1978) for a deeper insight into the mentioned constructions.

2.1 Newtonian Mechanics

Consider a particle of mass m moving in $\mathbb{R}^n$. The position of a particle $x = (x_1, ..., x_n)$ is a vector in $\mathbb{R}^n$. More precisely, $x(t) = (x_1(t), ..., x_n(t))$ is the position of the particle at time t. Let $v(t)$ and $a(t)$ denote the velocity and the acceleration at time t respectively. Then

$$v(t) = \dot{x}(t) = (\dot{x}_1(t), ..., \dot{x}_n(t)), \tag{2.1.1}$$
$$a(t) = \ddot{x}(t) = (\ddot{x}_1(t), ..., \ddot{x}_n(t)), \tag{2.1.2}$$

where $\dot{x}_i(t) = \frac{d}{dt}x_i(t)$ and $\ddot{x}_i(t) = \frac{d}{dt}\dot{x}_i(t) = \frac{d^2}{dt^2}x_i(t)$. We recall *Newton's second law* of motion:

$$F(x(t), \dot{x}(t)) = m\ddot{x}(t), \tag{2.1.3}$$

where F is a force acting on the particle with mass m depending on its position $x(t)$ and its velocity $\dot{x}(t)$. Hence, the trajectories of motion are given by solutions of (2.1.3). We note that (2.1.3) is a system of ordinary second-order differential equations and is non-linear in general.[1]

---

[1] Note that the nonlinearity condition depends on the nature of the function F.

© The Author(s), under exclusive license to Springer Nature Singapore Pte Ltd. 2023
N. Moshayedi, *Quantum Field Theory and Functional Integrals*,
SpringerBriefs in Physics, https://doi.org/10.1007/978-981-99-3530-7_2

Example 2.1.0.1 (The free particle on $\mathbb{R}^n$). In the case of a free particle moving in Euclidean space $\mathbb{R}^n$, the force becomes $F = 0$, which implies that (2.1.3) becomes $\ddot{x} = 0$, hence the trajectories of motion are given by $x(t) = at + b$ with $a, b \in \mathbb{R}^n$.

Example 2.1.0.2 (Harmonic oscillator in one dimension ($n = 1$)). For a harmonic oscillator on $\mathbb{R}$, the force is given by $F = -Kx$, where $K = \omega^2 m$ is the so-called *spring constant*. Here we have denoted by ω the angular velocity of the system. This is called *Hooke's law*. The equation of motion becomes $m\ddot{x} + Kx = 0$. Hence, the trajectories of motion are given by

$$x(t) = a\cos(\omega t) + b\sin(\omega t), \tag{2.1.4}$$

with $a, b \in \mathbb{R}$.

Usually, in Newtonian mechanics, we are interested in *solving* the *equations of motion* (2.1.3). One way to solve (2.1.3) would be to find conserved quantities which may help simplify the mathematical problem.

2.1.1 Conservation of Energy

Assume that the force F depends only on the position and has the form $F = -\nabla V(x)$, where $V \in C^\infty(\mathbb{R}^n)$ is some function. Such a force F is called a *conservative* force and V is called the *potential* of F. Since (2.1.3) is a second-order differential equation, the state space or *phase space* of (2.1.3) is $\mathbb{R}^{2n} = \{(x, v) \mid x, v \in \mathbb{R}^n\}$. We define the *total energy function* E by

$$E(x, v) := \frac{1}{2}m\|v\|^2 + V(x), \tag{2.1.5}$$

where $\|v\|^2 = \langle v, v \rangle$ with the standard inner product $\langle \cdot, \cdot \rangle$ on $\mathbb{R}^n$. The main significance of the total energy function is that it is *conserved*, meaning that its value along any trajectory of motion is *constant*.

Proposition 2.1.1.1 *Consider a particle moving in Euclidean space $\mathbb{R}^n$ satisfying Newton's second law of motion, i.e., Eq. (2.1.3). Then the total energy E changes with time according to the differential equation*

$$\frac{\mathrm{d}}{\mathrm{d}t}E(x(t), \dot{x}(t)) = 0, \tag{2.1.6}$$

along any trajectory $x(t)$ satisfying (2.1.3).

Proof Note that along a solution $x(t)$ of (2.1.3) we have

$$\frac{\mathrm{d}}{\mathrm{d}t}E(x,v) = \sum_{i=1}^{n}\frac{\partial E}{\partial x_i}\dot{x}_i + \sum_{i=1}^{n}\frac{\partial E}{\partial v_i}\dot{v}_i$$

$$= \sum_{i=1}^{n}\frac{\partial V}{\partial x_i}v_i + m\sum_{i=1}^{n}v_i\dot{v}_i \tag{2.1.7}$$

$$= (\nabla V + ma)v$$

$$= (-F + ma)v$$

$$= 0,$$

$\square$

Definition 2.1.1.1 (*Constant of motion*). Let f be a function on the phase space $\mathbb{R}^{2n}$. We say f is a *constant of motion* if $\frac{\mathrm{d}}{\mathrm{d}t}f = 0$ along $(x(t), \dot{x}(t))$, whenever $x(t)$ is a trajectory of motion.

Remark 2.1.1.1 Constants of motion are *conserved* quantities.

By Proposition 2.1.1.1, the total energy E is a constant of motion. By looking at an example, we can observe that the conservation of energy actually helps us to understand the solutions of the equations of motion. Let us first rewrite (2.1.3) in terms of first-order differential equations

$$\frac{\mathrm{d}}{\mathrm{d}t}x_i(t) = v_i(t), \quad i = 1, 2, \dots, n,$$

$$\frac{\mathrm{d}}{\mathrm{d}t}v_i(t) = \frac{1}{m}F_i(x(t)), \quad i = 1, 2, \dots, n. \tag{2.1.8}$$

For simplicity, assume $n = 1$. Hence, we have

$$\frac{\mathrm{d}}{\mathrm{d}t}x(t) = v(t),$$

$$\frac{\mathrm{d}}{\mathrm{d}t}v(t) = \frac{1}{m}F(x(t)). \tag{2.1.9}$$

By conservation of energy, we know that $\frac{\mathrm{d}}{\mathrm{d}t}E(x,v) = 0$ along $(x(t), v(t))$, whenever $(x(t), v(t))$ satisfies (2.1.9). Let $E(x(t), v(t)) = E_0$. Then, we get that

$$\frac{1}{2}m\dot{x}(t)^2 + V(x(t)) = E_0, \tag{2.1.10}$$

and thus

$$\dot{x}(t) = \pm\sqrt{\frac{2(E_0 - V(x(t)))}{m}}, \tag{2.1.11}$$

which can be solved using the method of *separation of variables*. From this example, we can see that the conservation of energy helps us to simplify the given system of equations in the 1-dimensional case, which was the previous example, and hence we were able to reduce the second-order differential equation to a first-order differential equation and even solve it. A general "mantra" in this setting is: *the knowledge of conserved quantities helps us to simplify the equations of motion.*

2.2 Hamiltonian Mechanics

2.2.1 The General Formulation

Classical mechanics is usually considered in two different settings besides Newtonian mechanics, namely *Lagrangian mechanics* and *Hamiltonian mechanics*. Hamiltonian mechanics gives a systematic approach to understanding conserved quantities. Consider a particle moving in Euclidean space $\mathbb{R}^n$. We now want to think of the total energy as a function of *position* and *momentum* rather than a function of position and velocity, i.e., we consider a Hamiltonian function

$$H(x, p) = \frac{1}{2m} \sum_{j=1}^{n} p_j^2 + V(x), \qquad (2.2.1)$$

where $p_j = m\dot{x}_j$ and $V \in C^\infty(\mathbb{R}^n)$ is some potential. The system of equations given in (2.2.1) can then be rewritten as

$$\begin{aligned}
\frac{\mathrm{d}}{\mathrm{d}t} x_i(t) &= x_i(t) = \frac{1}{m} p_i = \frac{\partial H}{\partial p_i}, \\
\frac{\mathrm{d}}{\mathrm{d}t} p_i(t) &= m \frac{\mathrm{d}}{\mathrm{d}t} x_i(t) = -\frac{\partial V}{\partial x_i} = -\frac{\partial H}{\partial x_i}.
\end{aligned} \qquad (2.2.2)$$

The equations of (2.2.2), i.e.,

$$\dot{x}_i = \frac{\partial H}{\partial p_i}, \qquad \dot{p}_i = -\frac{\partial H}{\partial x_i} \qquad (2.2.3)$$

are usually called *Hamilton's equations* or *canonical equations*. These are the equations of motion in the setting of Hamiltonian mechanics. Note that they are considerably easier than in Newtonian mechanics since they are first-order instead of second-order differential equations.

2.2.2 The Poisson Bracket

The previous observation implies that in Hamiltonian mechanics we consider the phase space to be

$$\mathbb{R}^{2n} := \{(x, p) \mid x, p \in \mathbb{R}^n\}.$$

It turns out that $\mathbb{R}^{2n}$ actually carries more structure than $\mathbb{R}^n$. For two functions $f, g \in C^\infty(\mathbb{R}^{2n})$, one can define the *Poisson bracket*

$$\{f, g\} := \sum_{j=1}^{n} \left(\frac{\partial f}{\partial x_j} \frac{\partial g}{\partial p_j} - \frac{\partial g}{\partial x_j} \frac{\partial f}{\partial p_j} \right). \tag{2.2.4}$$

Exercise 2.2.2.1 Show that the Poisson bracket satisfies the following properties for $f, g, h \in C^\infty(\mathbb{R}^{2n})$:

(1) $\{f, g\} = -\{g, f\}$, (*anti-symmetry*);
(2) $\{f, g + ch\} = \{f, g\} + c\{f, h\}$, $\forall c \in \mathbb{R}$, (*linearity*);
(3) $\{f, gh\} = \{f, g\}h + \{f, h\}g$, (*Leibniz rule*);
(4) $\{f, \{g, h\}\} = \{\{f, g\}, h\} + \{g, \{f, h\}\}$ (*Jacobi identity*).

Example 2.2.2.1 Let p_j and x_j be momentum and position *observables* as images of the following maps respectively:

$$\begin{aligned} (x, p) &\longmapsto p_j, \\ (x, p) &\longmapsto x_j. \end{aligned} \tag{2.2.5}$$

Then $\{x_i, x_j\} = 0 = \{p_i, p_j\}$ and $\{x_i, p_j\} = \delta_{ij}$, where δ_{ij} denotes the *Kronecker delta* defined as

$$\delta_{ij} = \begin{cases} 1, & i = j, \\ 0, & i \neq j. \end{cases}$$

The Poisson bracket can in fact be used to describe conserved quantities by considering the dynamics of the corresponding observable. To understand this, we first need the following proposition.

Proposition 2.2.2.1 *Time-evolution for an observable $f \in C^\infty(\mathbb{R}^{2n})$ is given by the equation*

$$\frac{d}{dt} f = \{f, H\}, \tag{2.2.6}$$

where $H \in C^\infty(\mathbb{R}^{2n})$ is the total energy, along a solution of Hamilton's equations $\{(x(t), p(t))\} \subset \mathbb{R}^{2n}$.

Exercise 2.2.2.2 Prove Proposition 2.2.2.1.

Corollary 2.2.2.1 *An observable $f \in C^\infty(\mathbb{R}^{2n})$ is conserved along solutions of Hamilton's equations if and only if*

$$\{f, H\} = 0.$$

Proof By Proposition 2.2.2.1, the evolution equation $\frac{d}{dt} f = \{f, H\}$ holds along solutions $(x(t), p(t))$ of Hamilton's equations. By definition, f is conserved if $\frac{d}{dt} f = 0$, which holds if and only if $\{f, H\} = 0$. $\qquad\qquad\qquad\qquad\qquad\qquad\qquad\qquad\square$

Remark 2.2.2.1 Given any observable $f \in C^\infty(\mathbb{R}^{2n})$, we can define Hamilton's equations by

$$\begin{aligned}
\dot{x}_i &= \frac{\partial f}{\partial p_i}, \\
\dot{p}_i &= -\frac{\partial f}{\partial x_i},
\end{aligned} \tag{2.2.7}$$

for all $i = 1, 2, ..., n$.

For everything that will follow now, we will assume that the reader is familiar with the basic notions of differential geometry.

Remark 2.2.2.2 The Euclidean space $\mathbb{R}^{2n}$ has a canonical symplectic structure given by the 2-form

$$\omega = \sum_{i=1}^{n} dp_i \wedge dx_i.$$

Given $f \in C^\infty(\mathbb{R}^{2n})$, there exists a vector field X_f, called the *Hamiltonian vector field of f*, defined by

$$\omega(X_f, \cdot) = -df \tag{2.2.8}$$

The flow of X_f is given by solutions of (2.2.7). In this case, one can check that

$$\{f, g\} = \omega(X_f, X_g). \tag{2.2.9}$$

This means that if (N, ω) is a symplectic manifold, then we can define the Poisson bracket of $f, g \in C^\infty(N)$ using (2.2.9).

Remark 2.2.2.3 Let $f \in C^\infty(\mathbb{R}^{2n})$ and X_f be the corresponding Hamiltonian vector field. The flow of X_f (or in other words the solutions of (2.2.7)) defines one-parameter diffeomorphisms

$$\begin{aligned}
\Phi_{X_f}^t &: \mathbb{R}^{2n} \longrightarrow \mathbb{R}^{2n}, \\
(x, p) &\longmapsto \Phi_{X_f}^t(x, p) = (x(t), p(t)),
\end{aligned} \tag{2.2.10}$$

where $x(t)$ and $p(t)$ satisfy Hamilton's equations with $x(0) = x$ and $p(0) = p$. Then, assuming that the flow is complete, we get

(1) $\Phi_{X_f}^t$ preserves ω (i.e., $(\Phi_{X_f}^t)^*\omega = \omega$). Such a map is called a *symplectomorphism*.

(2) $\Phi_{X_f}^t$ preserves the volume form

$$\mathrm{dVol} = \mathrm{d}x_1 \wedge \cdots \wedge \mathrm{d}x_n \wedge \mathrm{d}p_1 \wedge \cdots \wedge \mathrm{d}p_n,$$

i.e., $(\Phi_{X_f}^t)^*\mathrm{dVol} = \mathrm{dVol}$. This is known as *Liouville's theorem*.

Remark 2.2.2.4 Let $f, g \in C^\infty(\mathbb{R}^{2n})$. Then f is conserved along the solutions of Hamilton's equations of g if and only if $\{f, g\} = 0$. This is an instance of *Noether's theorem*.

2.3 Lagrangian Mechanics

Lagrangian mechanics gives a counterpart to Hamiltonian mechanics, where one considers methods of variational calculus in order to understand the behavior of classical mechanical systems. There are two important points in this formalism:

- Mechanics on a configuration space.
- Basic theorems are invariant under actions of diffeomorphisms of the configuration space. It is useful to compute conserved quantities.

2.3.1 Lagrangian System

Let M be a smooth manifold (we will usually consider $M = \mathbb{R}^n$). A *Lagrangian system* with configuration space M consists of a smooth real-valued function $L: TM \times \mathbb{R} \to \mathbb{R}$, where TM denotes the tangent bundle of M (e.g., if $M = \mathbb{R}^n$, then $T\mathbb{R}^n = \mathbb{R}^n \times \mathbb{R}^n = \{(x, v) \mid x \in \mathbb{R}^n, v \in \mathbb{R}^n\}$). The function L is called the *Lagrangian function* or simply *Lagrangian*. Lagrangian mechanics uses special ideas such as the *least action principle* from calculus of variation. Let $x_0, x_1 \in M$ and

$$P(M, x_0, x_1) := \{\gamma: \mathbb{R} \supset [t_0, t_1] \to M \mid \gamma(t_0) = x_0, \gamma(t_1) = x_1\},$$

which is the space of parametrized paths joining x_0 to x_1.

Definition 2.3.1.1 (*Action functional*). The *action functional* $S: P(M, x_0, x_1) \to \mathbb{R}$ of the Lagrangian system (M, L) is defined by

$$S[\gamma] = \int_{t_0}^{t_1} L(\gamma(t), \dot{\gamma}(t), t)\mathrm{d}t. \tag{2.3.1}$$

From now on we take $M = \mathbb{R}^n$. We are interested in understanding the *critical points* of the functional S. Let $h: [t_0, t_1] \to \mathbb{R}^n$ be such that $\gamma + h \in P(\mathbb{R}^n, x_0, x_1)$ and $h(t_0) = h(t_1) = 0$. We think of h as a small variation of $\gamma \in P(\mathbb{R}^n, x_0, x_1)$. Then, if we change $\gamma(t)$ by h, we get

$$S[\gamma + \varepsilon h] = \int_{t_0}^{t_1} L(\gamma(t) + \varepsilon h(t), \dot{\gamma}(t) + \varepsilon h(t), t)dt, \qquad (2.3.2)$$

which needs to be extremal with respect to the parameter ε. Hence, we get

$$\frac{d}{d\varepsilon} S[\gamma + \varepsilon h] = \int_{t_0}^{t_1} \left(\frac{\partial L}{\partial \gamma} h + \frac{\partial L}{\partial \dot{\gamma}} \dot{h} \right) dt = 0. \qquad (2.3.3)$$

For the second part, we use integration by parts, which gives

$$\int_{t_0}^{t_1} \frac{\partial L}{\partial \dot{\gamma}} \dot{h}(t)dt = \underbrace{\frac{\partial L}{\partial \dot{\gamma}} h \Big|_{t_0}^{t_1}}_{=0} - \int_{t_0}^{t_1} \frac{d}{dt} \frac{\partial L}{\partial \dot{\gamma}} h(t)dt. \qquad (2.3.4)$$

The last term remains and by the product rule we get

$$\int_{t_0}^{t_1} \left(\frac{\partial L}{\partial \gamma} - \frac{d}{dt} \frac{\partial L}{\partial \dot{\gamma}} \right) h(t)dt = 0. \qquad (2.3.5)$$

Definition 2.3.1.2 (*Extremal point/Critical point*). An *extremal (or critical) point* of S is some $x \in P(\mathbb{R}^n, x_0, x_1)$ such that

$$\int_{t_0}^{t_1} \left(\frac{\partial L}{\partial x} - \frac{d}{dt} \frac{\partial L}{\partial \dot{x}} \right) h dt = 0 \qquad (2.3.6)$$

along x for all paths h such that $h(t_0) = h(t_1) = 0$.

Theorem 2.3.1.1 *A path $x \in P(\mathbb{R}^n, x_0, x_1)$ is an extremal of S if and only if along x we have*

$$\frac{\partial L}{\partial x} - \frac{d}{dt} \frac{\partial L}{\partial \dot{x}} = 0 \qquad (2.3.7)$$

The proof for this theorem follows from the following lemma.

Lemma 2.3.1.1 *Let $f: [t_0, t_1] \to \mathbb{R}^n$ be a continuous path and*

$$\int_{t_0}^{t_1} f h dt = 0 \qquad (2.3.8)$$

for all continuous paths $h: [t_0, t_1] \to \mathbb{R}^n$ such that $h(t_0) = h(t_1) = 0$. Then $f \equiv 0$ on $[t_0, t_1]$.

Proof For simplicity assume $n = 1$, i.e., $f: [t_0, t_1] \to \mathbb{R}$ and $h: [t_0, t_1] \to \mathbb{R}$. By contradiction assume there is some $t \in [t_0, t_1]$ such that $f(t) > 0$. Then by continuity there is some $\delta > 0$ such that $f > 0$ on $(t - \delta, t + \delta)$. Let h be a continuous function on $[t_0, t_1]$ such that h vanishes outside $(t - \delta, t + \delta)$ but $h > 0$ on $(t - \delta/2, t + \delta/2)$. Then

$$\int_{t_0}^{t_1} fh\, dt > 0,$$

which is a contradiction. □

Definition 2.3.1.3 (*Euler–Lagrange equations*). The equations

$$\frac{\partial L}{\partial x} - \frac{\mathrm{d}}{\mathrm{d}t}\frac{\partial L}{\partial \dot{x}} = 0 \qquad (2.3.9)$$

are called the *Euler–Lagrange (EL) equations* of $S[x]$.

Corollary 2.3.1.1 *A path* $x \in P(\mathbb{R}^n, x_0, x_1)$ *is an extremal of* S *if and only if it satisfies the Euler–Lagrange equations.*

2.3.2 Hamilton's Least Action Principle

Recall that we defined the total energy function by

$$E(x, v) = \frac{1}{2}m\|v\|^2 + V(x),$$

where the first term is the kinetic energy and the second term is the potential energy.

Theorem 2.3.2.1 *Define* $L(\gamma(t), \dot{\gamma}(t), t) = \frac{1}{2}m\|\dot{\gamma}(t)\|^2 - V(\gamma(t))$. *Then an extremal path* $\gamma(t)$ *of* S *solves the system (2.1.8).*

Exercise 2.3.2.1 Prove Theorem 2.3.2.1.

Remark 2.3.2.1 Even though only an extremal path of S is involved here, it is called *Hamilton's least action principle.* Even though it is called least action principle, it actually considers the *stationary* paths. Hence, more correctly it should be called *principle of stationary action.*

Next, we want to briefly investigate how Hamilton's equations and the EL equations are related.

2.4 The Legendre Transform

We can ask about the relationship between the Hamiltonian and Lagrangian functions, in particular, how we can pass from one to the other. In fact, it turns out that they are related by the concept of a *Legendre transform*. Let f be a smooth convex function, i.e., its second derivative is positive $f''(x) > 0$. Let $p \in \mathbb{R}$ and define $g(x) := px - f(x)$. Then it is easy to see that $g'(x) = p - f'(x)$. Moreover, since f is convex, which means that its first derivative f' is increasing, there is a unique x_0 such that $g(x_0) = 0$. We denote this x_0 by $x(p)$. Moreover, the fact that the second derivative of f is positive, i.e., $f''(x) > 0$, we get that the second derivative of g is negative, i.e., $g''(x) < 0$, and hence g has a maximum at $x(p)$. In this case the *Legendre transform* of f, denoted by $\mathcal{L}f$, is defined by

$$\mathcal{L}f(p) := \max_x g(x) = \max_x (px - f(x)). \tag{2.4.1}$$

Example 2.4.0.1 Let $f(x) = x^2$, then $\mathcal{L}f(p) = \frac{1}{4}p^2$.

Example 2.4.0.2 Let $f(x) = \frac{1}{2}x^2$, then $\mathcal{L}f(p) = \frac{1}{2}p^2$.

More generally, let V be a finite-dimensional vector space with dual V^* and $f : V \to \mathbb{R}$ be a function. Then $\mathcal{L}f : V^* \to \mathbb{R}$ is defined by

$$\mathcal{L}f(p) := \max_{x \in V}(p(x) - f(x)), \tag{2.4.2}$$

where $p(x)$ is the pairing between $x \in V$ and $p \in V^*$. Actually, if f is convex, then $\mathcal{L}f$ exists.

Exercise 2.4.0.1 Show that if f is convex, then so is $\mathcal{L}f$. Moreover, show that the Legendre transform $\mathcal{L}$ is an involution, i.e., $\mathcal{L}(\mathcal{L}f) = f$.

Example 2.4.0.3 Let A be an $n \times n$ positive-definite matrix and consider the map $f : \mathbb{R}^n \to \mathbb{R}$, $f(x) = \frac{1}{2}\langle Ax, x \rangle$, where $\langle \cdot, \cdot \rangle$ is the standard inner product on $\mathbb{R}^n$. Then

$$\mathcal{L}f(\omega) = \frac{1}{2}\langle A^{-1}\omega, \omega \rangle.$$

Let us now consider a Lagrangian system $(\mathbb{R}^n, L)$, i.e., $L : \underbrace{\mathbb{R}^n \times \mathbb{R}^n \times \mathbb{R}}_{\ni (x, v, t)} \to \mathbb{R}$.

Moreover, let $H(x, p, t) = \mathcal{L}L$ be the Legendre transform of L in v-direction.

Theorem 2.4.0.1 *The system of EL equations are equivalent to Hamilton's equation with H defined as above.*

Exercise 2.4.0.2 Prove Theorem 2.4.0.1.

Chapter 3
The Schrödinger Picture of Quantum Mechanics

It is an interesting observation in nature that the theory of classical mechanics does not work for the scales of atoms and molecules. Consider the example of the hydrogen atom which is composed of two particles, namely a proton of charge[1] $+e$ and an electron of charge $-e$. If we follow the rules of classical mechanics, the charged electron would radiate energy continuously causing the atom to collapse. But this does not actually coincide with the observation and thus cannot be true. Hence, we need another theory that is consistent with such observations. *Quantum mechanics* is the actual candidate for explaining the stability of molecules and atoms.[2] Mathematically, it is a fascinating theory in itself. In this chapter we want to explore the mathematics behind some of its statements.

3.1 Postulates of Quantum Mechanics

One of the most important things in quantum mechanics is the notion of a *Hilbert space*. A *Hilbert space* $(\mathcal{H}, \langle \cdot, \cdot \rangle)$ is a complete inner product space where the norm for completeness is induced by the inner product, i.e. $\| \cdot \|^2 = \langle \cdot, \cdot \rangle$. We will not indicate the inner product when talking about a Hilbert space whenever it is clear. The *state space* in quantum mechanics is given by a Hilbert space. Let us now consider the main postulates of quantum mechanics. We will sometimes denote the inner product and the norm coming from a Hilbert space $\mathcal{H}$ by $\langle \cdot, \cdot \rangle_\mathcal{H}$ and $\| \cdot \|_\mathcal{H}$ respectively whenever it is not clear, in order to avoid any confusion with other inner products and norms.

---

[1] Here $e = 1.6 \times 10^{-19}$C denotes the elementary charge.

[2] We refer to a standard physics textbook on quantum mechanics for the actual motivation that led to the postulates of quantum mechanics.

© The Author(s), under exclusive license to Springer Nature Singapore Pte Ltd. 2023
N. Moshayedi, *Quantum Field Theory and Functional Integrals*,
SpringerBriefs in Physics, https://doi.org/10.1007/978-981-99-3530-7_3

3.1.1 First Postulate

The *pure states* of a quantum mechanical system are given by *rays* in a Hilbert space $\mathcal{H}$, i.e. 1-dimensional subspaces of $\mathcal{H}$. The Hilbert space $\mathcal{H}$ is called the *space of states*. Moreover, define the space

$$P\mathcal{H} := (\mathcal{H} \setminus \{0\})/(\mathbb{C} \setminus \{0\}).$$

Let $\phi, \psi \in \mathcal{H} \setminus \{0\}$. We say ϕ and ψ are *equivalent*, and we write $\phi \sim \psi$, if and only if there is an $\alpha \in \mathbb{C} \setminus \{0\}$ such that $\phi = \alpha\psi$. Then $P\mathcal{H}$ is the set of equivalence classes with respect to this equivalence relation.

Lemma 3.1.1.1 *There is a canonical bijection*

$$\left\{ 1\text{-dimensional subspaces of } \mathcal{H} \right\} \longleftrightarrow P\mathcal{H}.$$

Proof Let L be a 1-dimensional subspace of $\mathcal{H}$, and $\phi \in L$ such that $\phi \neq 0$. Define

$$\beta(L) = [\phi], \qquad [\phi] \in P\mathcal{H}.$$

We want to check that β is well-defined. Let $\psi \in L \setminus \{0\}$. Then there is some $\alpha \in \mathbb{C} \setminus \{0\}$ such that $\psi = \alpha\phi$ (since L is a 1-dimensional subspace). Thus, we get that $[\psi] = [\phi]$. This shows that β is well defined. It is then easy to check that β is a bijection. We leave this part to the reader. $\qquad\square$

Remark 3.1.1.1 The space $P\mathcal{H}$ is actually called the *space of pure states*.

From now on when we speak of a state we mean an element $\psi \in \mathcal{H}$ such that $\|\psi\|_{\mathcal{H}} = 1$, which are usually called *normalized states*. The concept of a state as a ray in $\mathcal{H}$ leads to the probabilistic interpretation of quantum mechanics. More precisely, this means that if a physical system is in the state ψ, then the probability that it is in the state ϕ is $|\langle \psi, \phi \rangle_{\mathcal{H}}|^2$. Since we assume that $\|\phi\|_{\mathcal{H}} = 1$ and $\|\psi\|_{\mathcal{H}} = 1$, clearly $0 \leq |\langle \psi, \phi \rangle_{\mathcal{H}}|^2 \leq 1$. Thus, it indeed makes sense to interpret the quantity $|\langle \psi, \phi \rangle_{\mathcal{H}}|^2$ as a probability.

3.1.2 Second Postulate

Beside the space of states, we also want to understand what quantum observables are. Quantum mechanical observables are *self-adjoint* operators on the space of states $\mathcal{H}$. Let A be an observable, i.e. a self-adjoint operator on $\mathcal{H}$. Then the *expectation* of A in the state $\psi \in \mathcal{H}$ is defined as

$$\langle A \rangle_\psi := \frac{\langle A\psi, \psi \rangle_{\mathcal{H}}}{\langle \psi, \psi \rangle_{\mathcal{H}}}. \tag{3.1.1}$$

We will denote the space of operators on a Hilbert space $\mathcal{H}$ by

$$\text{End}(\mathcal{H}) := \{A \colon \mathcal{H} \to \mathcal{H} \mid A \text{ linear}\}.$$

3.1.3 Third Postulate

Another question concerns the dynamics of a quantum mechanical system. More precisely, we want to understand how the general equation for the dynamics look like, either when considering states or observables. The dynamics are encoded in the Hamiltonian operator $\hat{H} \in \text{End}(\mathcal{H})$, which is the infinitesimal generator of the *unitary group* $U(t) = e^{-\frac{i}{\hbar}t\hat{H}}$. It fully describes the dynamics of the system. Let $\psi \in \mathcal{H}$ be a state. Then we can describe time-evolution by the *Schrödinger equation (SE)* (Schrödinger 1926),

$$i\hbar \frac{d}{dt}\psi(t) = \hat{H}\psi(t). \tag{3.1.2}$$

This is a dynamical equation with respect to the states. Using an Ansatz for Eq. (3.1.2), we can extract a solution of the form $\psi(t) = e^{-\frac{i}{\hbar}t\hat{H}}\psi(0)$. The analogue equation in the case of observables is constructed in the *Heisenberg picture*. The dynamical equation takes the form

$$\frac{d}{dt}A(t) = \frac{i}{\hbar}[\hat{H}, A(t)], \tag{3.1.3}$$

where A is an observable and $[\cdot, \cdot]$ is the commutator of operators, defined by

$$[A, B] := AB - BA.$$

Lemma 3.1.3.1 *Let $\phi(t)$ and $\psi(t)$ be solutions of (3.1.2), such that $\phi(0) = \phi_0$ and $\psi(0) = \psi_0$. Then, we get that*

$$\langle \phi(t), \psi(t) \rangle_{\mathcal{H}} = \langle \phi_0, \psi_0 \rangle_{\mathcal{H}}, \qquad \forall t \in \mathbb{R}.$$

Proof Note that we have

$$\phi(t) = e^{-\frac{i}{\hbar}t\hat{H}}\phi(0),$$
$$\psi(t) = e^{-\frac{i}{\hbar}t\hat{H}}\psi(0).$$

The result follows since $e^{-\frac{i}{\hbar}t\hat{H}}$ is a unitary operator. $\qquad\qquad\square$

3.1.4 Summary of Classical and Quantum Mechanics

As we have seen, there are three main mathematical differences between the classical and quantum setting, namely the states, the observables and the description for the dynamics. The following table summarizes these main differences of classical and quantum mechanics.

	Classical Mechanics	**Quantum Mechanics**
State space	T^*M; the cotangent bundle of some manifold M	$P\mathcal{H}$; the space of pure states for some Hilbert space $\mathcal{H}$
Observables	$C^\infty(T^*M)$; smooth functions on the state space	$\mathrm{End}(\mathcal{H})$ self-adjoint operators on the state space $\mathcal{H}$ (or on $P\mathcal{H}$)
Dynamics	Described by Hamilton's equations associated with a Hamiltonian function $H \in C^\infty(T^*M)$: $\dot{q} = \frac{\partial H}{\partial p}, \dot{p} = -\frac{\partial H}{\partial q}$	Described either by the Schrödinger equation associated with a quantum Hamiltonian operator $\hat{H}$: $i\hbar \frac{\mathrm{d}}{\mathrm{d}t}\psi(t) = \hat{H}\psi(t)$, or by the Heisenberg equation associated with a quantum Hamiltonian operator $\hat{H}$: $\frac{\mathrm{d}}{\mathrm{d}t}A(t) = \frac{i}{\hbar}[\hat{H}, A(t)]$

3.2 Elements of Functional Analysis

In this section we want to describe the most important aspects of operator theory and functional analysis that are needed in order to understand the properties of the objects and notions in the setting of quantum mechanics.

3.2.1 Bounded Operators

Let $\mathcal{H}$ be a Hilbert space. We will always assume it is *separable*, i.e. that there exists a basis of $\mathcal{H}$. An operator on $\mathcal{H}$ is a pair $(A, D(A))$ where $D(A) \subset \mathcal{H}$ is a subspace of $\mathcal{H}$, called the *domain* of A, and $A : D(A) \to \mathcal{H}$ is a linear map. We can always assume that $D(A)$ is dense in $\mathcal{H}$.

Definition 3.2.1.1 (*Bounded operator*) A linear map $A : D(A) \to \mathcal{H}$ is called *bounded* if there exists some $\varepsilon > 0$ such that for all $\psi \in D(A)$ we get that

$$\|A\psi\|_{\mathcal{H}} \leq \varepsilon \|\psi\|_{\mathcal{H}}.$$

Otherwise, we say A is *unbounded*.

Remark 3.2.1.1 If A is unbounded, it can always be extended to a bounded operator $\tilde{A}: \mathcal{H} \to \mathcal{H}$. Hence, when we talk about a bounded operator, we always consider $A: \mathcal{H} \to \mathcal{H}$.

Let $A: \mathcal{H} \to \mathcal{H}$ be a bounded operator. Then there is a unique operator $A^*: \mathcal{H} \to \mathcal{H}$ such that

$$\langle \phi, A\psi \rangle_{\mathcal{H}} = \langle A^*\phi, \psi \rangle_{\mathcal{H}}, \quad \forall \phi, \psi \in \mathcal{H}.$$

Definition 3.2.1.2 (*Adjoint/self-adjoint operator*) We call A^* the *adjoint* of A. Moreover, a bounded operator $A: \mathcal{H} \to \mathcal{H}$ is called *self-adjoint* if $A^* = A$.

For a measure space $(\Omega, \mathcal{A}, \mu)$, we can define the space of *p-integrable functions* $\mathcal{L}^p(\Omega) := \{f: \Omega \to \mathbb{R} \mid \int_{\Omega} |f|^p d\mu < \infty\}$. If we consider the equivalence relation on $\mathcal{L}^p(\Omega)$ given by $f \sim g$ if and only if $f = g$ almost everywhere, we can consider the equivalence classes $L^p(\Omega) = \mathcal{L}^p(\Omega)/\sim$. We usually just write f instead of $[f]_\sim$. Moreover, we can define a norm on $L^p(\Omega)$ by

$$\|f\|_p := \left(\int_{\Omega} |f|^p d\mu \right)^{\frac{1}{p}}.$$

For $p = 2$ we can also define an inner product by

$$\langle f, g \rangle_{L^2} := \int_{\Omega} fg \, d\mu.$$

Moreover, it is easy to see that $\langle f, f \rangle_{L^2} = \|f\|_2^2$. One can actually show that $L^2(\Omega)$ is complete with respect to this norm. Hence, for any measure space $(\Omega, \mathcal{A}, \mu)$, the space $L^2(\Omega)$ is a Hilbert space with inner product $\langle \cdot, \cdot \rangle_{L^2}$. We usually call $L^2(\Omega)$ the space of *square-integrable functions* on Ω. For complex-valued functions $f: \Omega \to \mathbb{C}$, we will additionally take a conjugate for the inner product, i.e. we have $\langle f, g \rangle_{L^2} := \int_{\Omega} \bar{f} g \, d\mu$.

Example 3.2.1.1 Let $\mathcal{H} = L^2([0, 1])$ and consider the operator $X: \mathcal{H} \to \mathcal{H}$ given by multiplication, i.e. $(Xf)(x) = xf(x)$ for all $x \in [0, 1]$. Then, we get that

$$\|Xf\|_2^2 = \int_0^1 x^2 |f(x)|^2 dx \leq \int_0^1 |f(x)|^2 dx = \|f\|_2^2,$$

which implies that $\|Xf\|_2 \leq \|f\|_2 < \infty$ and thus X is bounded. Consider now two square-integrable functions $f, g \in L^2([0, 1])$. Then, we get that

$$\langle f, Xg \rangle_{L^2} = \int_0^1 \overline{f(x)} xg(x) dx = \int_0^1 \overline{xf(x)} g(x) dx = \langle Xf, g \rangle_{L^2}, \qquad (3.2.1)$$

and thus $X^* = X$, which means that X is self-adjoint.

3.2.2 Unbounded Operators

Example 3.2.2.1 Let $\mathcal{H} = L^2(\mathbb{R})$ and let X be the same multiplication operator as before and define its domain $D(X) = \{\phi \in L^2(\mathbb{R}) \mid x\phi(x) \in L^2(\mathbb{R})\}$. We claim that

(1) $D(X)$ is dense in $L^2(\mathbb{R})$.
(2) X is unbounded.

To show (1), let $\phi \in L^2(\mathbb{R})$ and define $\phi_n = \phi\chi_{[-n,n]}$, where $\chi_{[-n,n]}$ denotes the characteristic function defined by

$$\chi_{[-n,n]}(x) = \begin{cases} 1, & x \in [-n, n], \\ 0, & x \notin [-n, n]. \end{cases}$$

It is then clear that $x\phi_n \in L^2(\mathbb{R})$ and by Lebesgue's dominated convergence theorem we have $\phi_n \xrightarrow{n \to \infty} \phi$ in $L^2(\mathbb{R})$, which proves (1). To see that X is unbounded, consider the sequence of functions $\phi_n = \frac{1}{\sqrt{n}}\chi_{[0,n]}$. Then $\|\phi_n\|_2 = 1$ for all n, but

$$\|X\phi_n\|_2^2 = \frac{1}{n}\int_0^n x^2 dx = \frac{n^2}{3} \xrightarrow{n \to \infty} \infty.$$

Thus, X is unbounded, which proves (2).

3.2.3 Adjoint of an Unbounded Operator

Let A be an unbounded operator on $\mathcal{H}$ with domain $D(A)$. We define the domain of its adjoint as

$$D(A^*) = \{\phi \in \mathcal{H} \mid \langle \phi, A \cdot \rangle_{\mathcal{H}} \text{ is a bounded linear functional on } D(A^*)\}.$$

Using *Riesz's representation theorem*, one can show that if $\phi \in D(A^*)$, then there is a unique $\psi \in \mathcal{H}$ such that

$$\langle \psi, \chi \rangle_{\mathcal{H}} = \langle \phi, A\chi \rangle_{\mathcal{H}}, \quad \forall \chi \in D(A).$$

We can then just define $\psi := A^*\phi$.

Definition 3.2.3.1 (*Symmetric operator*) Let A be an unbounded operator with domain $D(A)$. We say A is *symmetric* if

$$\langle \phi, A\psi \rangle_{\mathcal{H}} = \langle A\phi, \psi \rangle_{\mathcal{H}}, \quad \forall \phi, \psi \in D(A).$$

Moreover, A is self-adjoint if $D(A) = D(A^*)$ and $A^*\phi = A\phi$.

Exercise 3.2.3.1 Show that if A is symmetric, then $D(A) \subseteq D(A^*)$. Hence, A is self-adjoint if and only if A is symmetric and $D(A) = D(A^*)$.

Exercise 3.2.3.2 Let $\mathcal{H} = L^2(\mathbb{R})$ and $V : \mathbb{R} \to \mathbb{R}$ be a measurable map. For $x \in \mathbb{R}$ we define the domain

$$D(V(x)) = \{\phi \in L^2(\mathbb{R}) \mid V(x)\phi(x) \in L^2(\mathbb{R})\}$$

for the operator

$$V(x) \colon D(V(x)) \longrightarrow L^2(\mathbb{R}),$$
$$\phi \longmapsto V(x)\phi. \tag{3.2.2}$$

Proposition 3.2.3.1 *The operator $V(x)$ is self-adjoint.*

Proof We need to check that $D(V(x))$ is dense in $L^2(\mathbb{R})$. It is easy to see that $V(x)$ is symmetric and thus $D(V(x)) = D(V(x)^*)$. Moreover, we can check that $D(V(x))$ is dense in $L^2(\mathbb{R})$. Since V is a real-valued function, $V(x)$ is symmetric as well. We only need to show that $D(V(x)^*) \subseteq D(V(x))$. To show this, let $\phi \in D(V(x)^*)$. We want to show that $V(x)\phi(x) \in L^2(\mathbb{R})$. Since $\phi \in D(V(x)^*)$, we get that $\psi \mapsto \langle \phi, V(x)\psi \rangle_{L^2}$ is a bounded linear functional on $D(V(x))$. In fact, it can be extended to a bounded linear functional on $L^2(\mathbb{R})$, since $D(V(x))$ is dense. Hence, by Riesz's representation theorem, there is a unique $\chi \in L^2(\mathbb{R})$ such that

$$\langle \chi, \psi \rangle_{L^2} = \langle \phi, V(x)\psi \rangle_{L^2}, \quad \forall \psi \in L^2(\mathbb{R}), \tag{3.2.3}$$

and thus

$$\int_{\mathbb{R}} \overline{\chi(x)}\psi(x)\mathrm{d}x = \int_{\mathbb{R}} \overline{\phi(x)}V(x)\psi(x)\mathrm{d}x, \quad \forall \psi \in L^2(\mathbb{R}). \tag{3.2.4}$$

Hence, we get that

$$\int_{\mathbb{R}} \overline{\chi(x)}\psi(x)\mathrm{d}x = \int_{\mathbb{R}} \overline{\phi(x)V(x)}\psi(x)\mathrm{d}x, \quad \forall \psi \in L^2(\mathbb{R}). \tag{3.2.5}$$

This shows that $\chi = V(x)\phi$ almost everywhere, and therefore $V(x)\phi \in L^2(\mathbb{R})$, which finally implies that $\phi \in D(V(x))$. $\qquad\square$

Similarly, one can show that the operator $\hat{p}$, defined by $\hat{p}\psi(x) := -i\hbar\frac{\mathrm{d}}{\mathrm{d}x}\psi(x)$, is a self-adjoint operator with domain

$$D(\hat{p}) = \{\psi \in L^2(\mathbb{R}) \mid k\hat{\psi}(k) \in L^2(\mathbb{R})\},$$

where

$$\hat{\psi}(k) = \frac{1}{2\pi} \int_{\mathbb{R}} e^{-ikx}\psi(x)\mathrm{d}x$$

is the *Fourier transform* of ψ. Next, we mention two technical results without a proof.

Theorem 3.2.3.1 (Spectral theorem/Functional calculus) *Let A be a self-adjoint operator on some Hilbert space* $\mathcal{H}$. *Let* $L(\mathcal{H})$ *denote the space of bounded linear operators on* $\mathcal{H}$. *Then, there is a unique map*

$$\hat{\phi}\colon \left\{ \text{Bounded measurable functions on } \mathbb{R} \right\} \longrightarrow L(\mathcal{H}),$$

such that

(1) $\hat{\phi}$ *is linear and* $\hat{\phi}(fg) = \hat{\phi}(f)\hat{\phi}(g)$ *for all bounded measurable functions* f, g *on* $\mathbb{R}$,
(2) $\hat{\phi}(f) = (\hat{\phi}(f))^*$, *for all bounded measurable functions* f *on* $\mathbb{R}$,
(3) $\|\hat{\phi}(f)\| \leq \|f\|_\infty$, *for all bounded measurable functions* f *on* $\mathbb{R}$, *where* $\|\cdot\|$ *denotes the* operator norm *defined by*

$$\|A\| := \sup_{\psi \in \mathcal{H}\setminus\{0\}} \frac{\|A\psi\|_\mathcal{H}}{\|\psi\|_\mathcal{H}}.$$

(4) If $f_n \xrightarrow{n\to\infty} x$ *and* $|f_n(x)| \leq |x|$, *where* f_n *are bounded measurable functions on* $\mathbb{R}$ *for all* $n \in \mathbb{N}$, *then for all* $\psi \in D(A)$ *we have*

$$\hat{\phi}(f_n)\psi \xrightarrow{n\to\infty} A\psi.$$

(5) $A\psi = \lambda\psi$ *for* $\lambda \in \mathbb{C}$.

We can use Theorem 3.2.3.1 to produce bounded operators from a self-adjoint operator, e.g. consider the function $f(x) = e^{itx}$. It is not hard to see that f is bounded and measurable. Hence, we can talk about $f(A) = e^{itA}$ as a bounded linear operator on $\mathcal{H}$.

Theorem 3.2.3.2 (Stone (1932)) *Let A be a self-adjoint operator on some Hilbert space* $\mathcal{H}$. *Define* $U(t) := e^{itA}$. *Then we get that:*

(1) $U(t)$ *is a unitary operator, i.e. we have*

$$\langle U(t)\phi, U(t)\psi \rangle = \langle \phi, \psi \rangle$$

for all $\phi, \psi \in \mathcal{H}$. *Moreover,* $U(t + s) = U(t) \circ U(s)$.
(2) for $\phi \in \mathcal{H}$ *we have that* $U(t)\phi \xrightarrow{t\to t_0} U(t_0)\phi$ *in* $\mathcal{H}$ *(strong convergence)*
(3) the limit $\lim_{t\to 0} \frac{U(t)\psi - \psi}{t}$ *exists in* $\mathcal{H}$ *for all* $\psi \in D(A)$ *and*

$$\lim_{t\to 0} \frac{U(t)\psi - \psi}{t} = iA\psi.$$

Formally, this means $\frac{\mathrm{d}}{\mathrm{d}t}U(t) = iA$.

(4) we have $\psi \in D(A)$ for all $\psi \in \mathcal{H}$ such that the limit $\lim\limits_{t \to 0} \frac{U(t)\psi - \psi}{t}$ exists.

Moreover, if $U(t)$, for $t \in \mathbb{R}$, is a family of unitary operators such that (1) and (2) hold, then $U(t) = e^{itA}$ for some self-adjoint operator A.

Definition 3.2.3.2 (*Strongly continuous one-parameter unitary group*) A family $U(t)$ satisfying (1) and (2) of Theorem 3.2.3.2 is called *strongly continuous one-parameter unitary group* and A is called the infinitesimal generator.

Definition 3.2.3.3 (*Resolvent*) Let A be an operator with domain $D(A)$, I the identity operator on $\mathcal{H}$ and $\lambda \in \mathbb{C}$. We say that A is in the *resolvent* set $\rho(A)$ of A if:

(1) $\lambda I - A \colon D(A) \to \mathcal{H}$ is bijective,
(2) $(\lambda I - A)^{-1}$ is a bounded operator.

Definition 3.2.3.4 (*Spectrum*) The spectrum $\sigma(A)$ of A is defined by

$$\sigma(A) := \mathbb{C} \setminus \rho(A).$$

One can actually check that if λ is an eigenvalue of A, then $\lambda \in \sigma(A)$. We call the set of eigenvalues of A the *point spectrum* of A.

Let $\mathcal{A}[0, 1]$ denote the set of absolutely continuous L^2-functions on $[0, 1]$.

Example 3.2.3.1 Consider the operator $T := i\frac{d}{dx}$ on $L^2([0, 1])$ with domain $D(T) := \mathcal{A}[0, 1]$. Then the spectrum of T is given by $\sigma(T) = \mathbb{C}$, which follows from a simple differential equation.

Example 3.2.3.2 Consider the operator $T := i\frac{d}{dx}$ with domain $D(T) := \{f \in \mathcal{A}[0, 1] \mid f(0) = 0\}$. We claim that the resolvent of T is given by $\rho(T) = \mathbb{C}$.

Proof of Example 3.2.3.2 Let $\lambda \in \mathbb{C}$ and define

$$S_\lambda g(x) := i \int_0^x e^{-i\lambda(x-s)} g(s) ds.$$

One can show that $(T - \lambda I)S_\lambda g = g$ for all $g \in L^2([0, 1])$. Moreover, $S_\lambda (T - \lambda I)g = g$ for all $g \in D(T)$. We need to show that S_λ is bounded. Indeed, we have

$$\|S_\lambda g\|_2^2 = \int_0^1 |S_\lambda g(x)|^2 dx \le \sup_{x \in [0,1]} |S_\lambda g(x)|^2 = \sup_{x \in [0,1]} \left| \int_0^x e^{-i\lambda(x-s)} g(s) ds \right|^2$$

$$\le \sup_{x \in [0,1]} \left(\int_0^x |e^{-i\lambda(x-s)} g(s)| ds \right)^2 \le \left(\sup_{x \in [0,1]} \left| \int_0^x e^{-i\lambda(x-s)} dx \right|^2 \right) \left(\sup_{x \in [0,1]} \left| \int_0^x g(s) ds \right|^2 \right)$$

$$\le C(\lambda) \|g\|_2^2.$$

$\square$

Theorem 3.2.3.3 *Let A be a self-adjoint operator on some Hilbert space $\mathcal{H}$. Then $\sigma(A) \subseteq \mathbb{R}$.*

Proof Assume A is bounded. Let $\lambda = a + ib$ with $b \neq 0$. We claim that $\lambda \in \rho(A)$. Let $\psi \in \mathcal{H}$. Moreover, define $T := (A - aI)$, where I is the identity operator on $\mathcal{H}$. Then

$$\langle (A - \lambda I)\psi, (A - \lambda I)\psi \rangle_{\mathcal{H}} = \langle (A - aI)\psi - ib\psi, (A - aI)\psi - ib\psi \rangle_{\mathcal{H}}$$
$$= \|T\psi\|_{\mathcal{H}}^2 - \langle ib\psi, T\psi \rangle_{\mathcal{H}} - \langle T\psi, ib\psi \rangle_{\mathcal{H}} + b^2 \|\psi\|_{\mathcal{H}}^2$$
$$= \|T\psi\|_{\mathcal{H}}^2 + b^2 \|\psi\|_{\mathcal{H}}^2$$
$$> b^2 \|\psi\|_{\mathcal{H}}^2.$$

Hence $\langle (A - \lambda I)^*(A - \lambda I)\psi, \psi \rangle_{\mathcal{H}} > b^2 \|\psi\|_{\mathcal{H}}^2$ and thus $(A - \lambda I)^*(A - \lambda I)$ is a positive operator. Moreover, we can show that $(A - \lambda I)^{-1}$ is bounded. $\qquad\square$

Remark 3.2.3.1 There are also plenty of examples for unbounded operators.

3.2.4 Sobolev Spaces

Consider an interval $I = (a, b) \subset \mathbb{R}$ and a differentiable function f with derivative f' being locally integrable on I. Moreover, let $\varphi \in C_c^\infty(I)$ be some *test function* with compact support. Then we get that

$$\int_I f'(t)\varphi(t)\mathrm{d}t = -\int_I f(t)\varphi'(t)\mathrm{d}t,$$

where we have used integration by parts. If we drop the integrability condition on the derivative f', we get that the integral on the left-hand side is not necessary well-defined. If the function f is itself locally integrable on I, it can happen that there is another locally integrable function g on I such that

$$\int_I g(t)\varphi(t)\mathrm{d}t = -\int_I f(t)\varphi'(t)\mathrm{d}t, \qquad \forall \varphi \in C_c^\infty(I).$$

Such a function g is called a *weak derivative* of f. Usually, it is also denoted by $f' := g$. This notion can be generalized to *higher weak derivatives*. Let $\Omega \subseteq \mathbb{R}^n$ and $f : \Omega \to \mathbb{R}$ be some locally integrable function on Ω. Moreover, let $\alpha = (\alpha_1, ..., \alpha_n) \in \mathbb{N}^n$ be a multi-index. A locally integrable function g on Ω is then called a *weak derivative of f of order* α, if for all test functions $\varphi \in C_c^\infty(\Omega)$ we get

$$\int_\Omega g(x)\varphi(x)\mathrm{d}x = (-1)^{|\alpha|} \int_\Omega f(x) D^\alpha \varphi(x)\mathrm{d}x.$$

Here we have $|\alpha| = \sum_{i=1}^n \alpha_i$ and $D^\alpha = \frac{\partial^{|\alpha|}}{\partial^{\alpha_1} x_1 \cdots \partial^{\alpha_n} x_n}$. We often just write $g = D^\alpha f$. It is actually possible to just assume that $f, g \in L^p(\Omega)$ for $1 \leq p \leq \infty$. The *Sobolev space* is then defined as the subset of $L^p(\Omega)$ in which $n \geq 1$ weak derivatives exist. In particular, it is given by the set

$$W^{k,p}(\Omega) := \left\{ u \in L^p(\Omega) \mid \forall \alpha \in \mathbb{N}^n \text{ with } |\alpha| \leq k \text{ there are weak derivatives } D^\alpha u \in L^p(\Omega) \right\}.$$

It is also possible to define a norm on the space $W^{k,p}(\Omega)$ which is called the *Sobolev norm*. It is defined as

$$\|u\|_{W^{k,p}(\Omega)} := \begin{cases} \left(\sum_{|\alpha| \leq k} \|D^\alpha u\|^p_{L^p(\Omega)} \right)^{1/p}, & p < \infty, \\ \max_{|\alpha| \leq k} \|D^\alpha u\|_{L^\infty(\Omega)}, & p = \infty. \end{cases}$$

Remark 3.2.4.1 For $p = 2$ we get that the Sobolev space is a Hilbert space and we write

$$H^k(\Omega) := W^{k,2}(\Omega).$$

3.3 Quantization of a Classical System

We want to talk about quantization of a classical system by considering a "*quantization map*" between the classical and quantum data. Consider a map $\mathscr{Q}$, the quantization map, which maps a classical system to a quantum system in the following sense: The classical space (phase space) of states (T^*M, ω), which is a symplectic manifold with symplectic structure given as the canonical symplectic structure on a cotangent bundle, is mapped to a Hilbert space $\mathcal{H}$. Moreover, the space of observables $C^\infty(T^*M)$ is mapped to the space of self-adjoint operators on $\mathcal{H}$. In particular, we know that the space of observables $C^\infty(T^*M)$ is endowed with a Poisson bracket $\{\cdot, \cdot\}$, which leads to the question of what its image is under the quantization map $\mathscr{Q}$.

Example 3.3.0.1 Consider $M = \mathbb{R}^n$ and its corresponding cotangent bundle $T^*M = \mathbb{R}^{2n} = \{(x, p) \mid x, p \in \mathbb{R}^n\}$. Then x_i and p^i represent classical position and momentum observables respectively and satisfy the commutation relation $\{x_i, p^j\} = \delta_{ij}$. Denote by $\hat{x}_i$ the operator given by multiplication with x_i and by $\hat{p}^i := -i\hbar \frac{\partial}{\partial x_i}$ the differential operator given by differentiation with respect to the position coordinate x_i (up to some constant). The operators $\hat{x}_i$ and $\hat{p}^i$ satisfy the commutation relation $[\hat{x}_i, \hat{p}^j] = i\hbar\delta_{ij}$, where $[\cdot, \cdot]$ again denotes the usual commutator of operators.

3.3.1 Definition

Example 3.3.0.1 can be generalized such that given $\{f, g\}$ for $f, g \in C^\infty(T^*M)$ it will be mapped by $\mathscr{Q}$ to

$$\frac{i}{\hbar}[\mathscr{Q}(f), \mathscr{Q}(g)],$$

or, by considering the classical equation for the dynamics of an observable $\frac{df}{dt} = \{f, H\}$, we get the quantum image

$$\frac{d}{dt}A(t) = \frac{i}{\hbar}[\hat{H}, A(t)],$$

which is the dynamical equation in the Heisenberg picture.

Definition 3.3.1.1 (*Quantization*) A *quantization* of a classical system $(\mathbb{R}^{2n}, \omega)$ is a tuple consisting of a Hilbert space $\mathcal{H}$ and a linear map

$$\mathscr{Q}: C^\infty(\mathbb{R}^{2n}) \to \{\text{self-adjoint operators on } \mathcal{H}\}$$

such that the following axioms hold:

(Q1) $\mathscr{Q}$ is linear,
(Q2) $\mathscr{Q}(\mathrm{id}_{\mathbb{R}^{2n}}) = \mathrm{id}_{\mathcal{H}}$,
(Q3) $\mathscr{Q}(x_i) = \hat{x}_i$, and $\mathscr{Q}(p_i) = \hat{p}_i$,
(Q4) $[\mathscr{Q}(f), \mathscr{Q}(g)] = i\hbar\mathscr{Q}(\{f, g\})$,
(Q5) $\mathscr{Q}(\phi \circ f) = \phi(\mathscr{Q}(f))$ for any map $\phi: \mathbb{R} \to \mathbb{R}$.

Remark 3.3.1.1 It turns out that it is actually very difficult to find such a map $\mathscr{Q}$. In fact, there is an actual problem when we impose (Q1)–(Q5) in general.

Example 3.3.1.1 For simplicity, let us consider the case when $n = 1$. We want to understand what the image of the classical observable $x^2 p^2$ is under such a quantization map $\mathscr{Q}$, i.e. we want to understand $\mathscr{Q}(x^2 p^2)$. For this, we use a trick and we write

$$x^2 p^2 = \frac{(x^2 + p^2)^2 - x^4 - p^4}{2}.$$

Then we use (Q3) and (Q5) to get the quantum observables

$$\mathscr{Q}(x^2 p^2) = \frac{(\hat{x}^2 + \hat{p}^2)^2 - \hat{x}^4 - \hat{p}^4}{2} = \frac{\hat{p}^2\hat{x}^2 + \hat{x}^2\hat{p}^2}{2}.$$

On the other hand, we have

$$xp = \frac{(x + p)^2 - x^2 - p^2}{2}.$$

Thus, we get

$$\mathscr{Q}(x^2 p^2) = \mathscr{Q}((xp)^2) = \left(\frac{(\hat{x}^2 + \hat{p}^2)^2 - \hat{x}^4 - \hat{p}^4}{2}\right)^2,$$

which implies

$$\mathcal{Q}(x^2 p^2) = \left(\frac{\hat{p}^2\hat{x}^2 + \hat{x}^2\hat{p}^2}{2}\right)^2,$$

but this result is different from what we had before.

The question here is: what are general approaches to resolve that issue? There are actually two different ways to solve this problem.

- Keep (Q1)–Q(4) and choose an appropriate domain for $\mathcal{Q}$,
- Keep (Q1)–(Q3) and assume that (Q4) holds asymptotically, i.e.

$$[\mathcal{Q}(f), \mathcal{Q}(g)] = i\hbar\mathcal{Q}(\{f, g\}) + O(\hbar^2).$$

There are two main approaches:

(1) (Canonical quantization) Here we quantize the observables x_i, p_i as the image of $\mathcal{Q}$, i.e. $x_i \mapsto \hat{x}_i$ and $p_i \mapsto \hat{p}_i$. Moreover, we want $f(x, p) \mapsto f(\hat{x}, \hat{p})$ and therefore we ask about the image of $x_i p_j$. In fact, there is an ordering problem. We need to make sure that we define $\mathcal{Q}(x_i^2 p_j^2)$ correctly.

(2) (Wick ordering quantization) Consider the complex numbers $z = x + i\alpha p$ and $\bar{z} = x - i\alpha p$ with $\alpha \in \mathbb{R}$. Then we can write $f(x, p)$ as $f(z, \bar{z})$, e.g. by setting

$$f(z, \bar{z}) = \sum_{ij} a_{ij} z_i^{r_i} \bar{z}_j^{r_j}, \quad a_{ij} \in \mathbb{C},$$

and by defining $\hat{z} := \hat{x} + i\alpha\hat{p}, \hat{\bar{z}} := \hat{z}^* = \hat{x} - i\alpha\hat{p}$ we can define

$$\mathcal{Q}_{\text{Wick}}(f) := f(\hat{z}, \hat{\bar{z}}) = \sum_{ij} \hat{z}_i^{r_i}(\hat{z}_j^{r_j})^*.$$

Example 3.3.1.2 Consider the case when $n = 1$. Then, by writing $x = \frac{1}{2}(z + \bar{z})$, we get

$$\begin{aligned}
\mathcal{Q}_{\text{Wick}}(x^2) &= \mathcal{Q}_{\text{Wick}}\left(\frac{1}{4}z^2 + \frac{1}{2}z\bar{z} + \frac{1}{4}\bar{z}^2\right) \\
&= \frac{1}{4}\left((\hat{x} + i\alpha\hat{p})^2 + 2(\hat{x} + i\alpha\hat{p})(\hat{x} - i\alpha\hat{p}) + (\hat{x} + i\alpha\hat{p})^2\right) \\
&= \frac{1}{4}\left((\hat{x}^2 - \alpha^2\hat{p}^2 + i\alpha(\hat{x}\hat{p} + \hat{p}\hat{x}) + 2(\hat{x}^2 + \alpha^2\hat{p}^2 + i\alpha[\hat{x}, \hat{p}]) + \hat{x}^2 - \alpha^2\hat{p}^2 - i\alpha(\hat{x}\hat{p} + \hat{p}\hat{x})\right) \\
&= \frac{1}{4}\left(4\hat{x}^2 + 2i\alpha[\hat{x}, \hat{p}]\right) = \hat{x}^2 - \frac{1}{2}\hbar\alpha I,
\end{aligned}$$

where I is the identity operator on $\mathcal{H}$.

(3) (Weyl quantization) Consider the case when $n = 1$. We define $\mathcal{Q}_{\text{Weyl}}(x, p) := \frac{\hat{x}\hat{p} + \hat{p}\hat{x}}{2}$, e.g. $\mathcal{Q}_{\text{Weyl}}(x^2 p) = \mathcal{Q}_{\text{Weyl}}(xxp) = \frac{\hat{x}^2\hat{p} + \hat{x}\hat{p}\hat{x} + \hat{p}\hat{x}^2}{3!}$. More generally, we can define

$$\mathscr{D}_{\text{Weyl}}(x_1 \cdots x_n p_1 \cdots p_m) = \frac{1}{(n+m)!} \sum_{\sigma \in S_{n+m}} \hat{x}_{\sigma(1)} \cdots \hat{x}_{\sigma(n)} \hat{p}_{\sigma(1)} \cdots \hat{p}_{\sigma(m)},$$

where S_k denotes the symmetric group of order k.

Exercise 3.3.1.1 Let g be any polynomial in x and p. Then we get that

$$\mathscr{D}_{\text{Weyl}}(xg) = \mathscr{D}_{\text{Weyl}}(x)\mathscr{D}_{\text{Weyl}}(g) - \frac{i\hbar}{2}\mathscr{D}_{\text{Weyl}}\left(\frac{\partial g}{\partial p}\right) = \mathscr{D}_{\text{Weyl}}(g)\mathscr{D}_{\text{Weyl}}(x) - \frac{i\hbar}{2}\mathscr{D}_{\text{Weyl}}\left(\frac{\partial g}{\partial p}\right),$$

$$\mathscr{D}_{\text{Weyl}}(pg) = \mathscr{D}_{\text{Weyl}}(p)\mathscr{D}_{\text{Weyl}}(g) + \frac{i\hbar}{2}\mathscr{D}_{\text{Weyl}}\left(\frac{\partial g}{\partial x}\right) = \mathscr{D}_{\text{Weyl}}(g)\mathscr{D}_{\text{Weyl}}(p) - \frac{i\hbar}{2}\mathscr{D}_{\text{Weyl}}\left(\frac{\partial g}{\partial x}\right).$$

Proposition 3.3.1.1 *Let f be a polynomial in x and p of degree at most 2 and g be any polynomial. Then we get that*

$$[\mathscr{D}_{\text{Weyl}}(f), \mathscr{D}_{\text{Weyl}}(g)] = i\hbar\mathscr{D}_{\text{Weyl}}(\{f, g\}).$$

Proof Consider the polynomial $f(x) = x$. Then we get that $\{f, g\} = \{x, g\} = \frac{\partial g}{\partial p}$. Using Exercise 3.3.1.1, we get that

$$[\mathscr{D}_{\text{Weyl}}(x), \mathscr{D}_{\text{Weyl}}(g)] = \frac{i\hbar}{2}\mathscr{D}_{\text{Weyl}}\left(\frac{\partial g}{\partial p}\right) + \frac{i\hbar}{2}\mathscr{D}_{\text{Weyl}}\left(\frac{\partial g}{\partial p}\right) = i\hbar\mathscr{D}_{\text{Weyl}}\left(\frac{\partial g}{\partial p}\right).$$

$\square$

Remark 3.3.1.2 This is actually not possible for arbitrary polynomials f and g. This result is due to a theorem of *Groenewold* (1946).

3.3.2   Eigenvalues of a Single Harmonic Oscillator

Consider the Hilbert space $\mathcal{H} = L^2(\mathbb{R})$ and the Hamiltonian $H(x, p) = \frac{1}{2m}p^2 + \frac{kx^2}{2}$ with $k = m\omega^2$ for some[3] $m > 0$ and $\omega \in \mathbb{R}$. Then going to the corresponding operator formulation, we have $\hat{p} = i\hbar\frac{d}{dx}$ and $\hat{x}$ is just multiplication by x. The Hamiltonian operator is then given by

$$\hat{H} = \frac{1}{2m}\hat{p}^2 + \frac{k\hat{x}^2}{2} = \frac{1}{2m}\left(\hat{p}^2 + (m\omega\hat{x})^2\right).$$

In the following we will only consider formal computations (i.e. we forget about the domains). Define the operators

---

[3] m denotes the *mass* of the particle and ω the *angular velocity* for the harmonic oscillator.

$$a := \frac{m\omega\hat{x} + i\hat{p}}{\sqrt{2\hbar m\omega}}, \tag{3.3.1}$$

$$a^* := \frac{m\omega\hat{x} - i\hat{p}}{\sqrt{2\hbar m\omega}}. \tag{3.3.2}$$

Lemma 3.3.2.1 *Let I be the identity operator on $\mathcal{H}$. Then we have*

$$\hat{H} = \hbar\omega\left(a^*a + \frac{1}{2}I\right).$$

Lemma 3.3.2.2 *Let I be the identity operator on $\mathcal{H}$. Then the following hold:*

(1) $[a, a^*] = I$,
(2) $[a, a^*a] = a$,
(3) $[a^*, a^*a] = -a^*$.

Exercise 3.3.2.1 Prove Lemma 3.3.2.2.

Proposition 3.3.2.1 *Assume that $\psi \in \mathcal{H}$ is an eigenvector of the operator a^*a with eigenvalue λ. Then we get that*

$$a^*a(a\psi) = (\lambda - 1)a\psi, \tag{3.3.3}$$

$$a^*a(a^*\psi) = (\lambda + 1)a^*\psi. \tag{3.3.4}$$

Remark 3.3.2.1 The consequence of Proposition 3.3.2.1 is that either $a\psi$ is an eigenvector or $a\psi = 0$. We know that $a^*a \geq 0$, so all eigenvalues are non-negative. Hence, if ψ is an eigenvector with eigenvalue λ, then there is some number $N \in \mathbb{N}$ such that $a^N\psi \neq 0$ but $a^{N+1}\psi = 0$.

Define $\psi_0 := a^N\psi$. Then $a^*a\psi_0 = 0$ and thus ψ_0 is an eigenvector of the operator a^*a with eigenvalue 0.

Proposition 3.3.2.2 *Let ψ_0 be such that $\|\psi_0\|_{\mathcal{H}} = 1$ and $a\psi_0 = 0$. Then, if we define $\psi_n := (a^*)^n\psi_0$ for $n \geq 0$, we get that:*

(i) $a^*\psi_n = \psi_{n+1}$,
(ii) $(a^*a)\psi_n = n\psi_n$,
(iii) $\langle\psi_n, \psi_m\rangle_{\mathcal{H}} = n!\delta_{mn}$,
(iv) $a\psi_{n+1} = (n+1)\psi_n$.

Remark 3.3.2.2 Next, we want to find some $\psi_0 \in L^2(\mathbb{R})$ such that $a\psi_0 = 0$ and $\|\psi_0\|_2^2 = 1$.

In order to do this, let us first define $\tilde{x} := \frac{x}{\sqrt{\frac{\hbar}{m\omega}}}$. Then we get that $\frac{d}{d\tilde{x}} = \sqrt{\frac{\hbar}{m\omega}}\frac{d}{dx}$ and thus

$$a = \frac{1}{\sqrt{2}}\left(\tilde{x} + \frac{d}{d\tilde{x}}\right), \qquad a^* = \frac{1}{\sqrt{2}}\left(\tilde{x} - \frac{d}{d\tilde{x}}\right).$$

Now, we want to solve the equation $a\psi_0 = 0$. This is then equivalent to the equation $\frac{d\psi_0}{d\tilde{x}} + \tilde{x}\psi_0 = 0$, which implies that

$$\psi_0(x) = \sqrt{\frac{2m\omega}{\hbar}} e^{-\frac{m\theta}{2\hbar}x^2} \in \mathcal{S}(\mathbb{R}).$$

Here $\mathcal{S}(\mathbb{R})$ represents the space of *Schwartz functions* on $\mathbb{R}$. In general, the space of Schwartz functions on $\mathbb{R}^n$ (with values in $\mathbb{C}$) is defined as

$$\mathcal{S}(\mathbb{R}^n, \mathbb{C}) := \{f \in C^\infty(\mathbb{R}^n, \mathbb{C}) \mid \forall \alpha, \beta \in \mathbb{N}^n, \ \|f\|_{\alpha,\beta} < \infty\},$$

where $\|f\|_{\alpha,\beta} := \sup\limits_{x \in \mathbb{R}^n} |x^\alpha D^\beta f(x)|$ with $D^\beta f = \frac{\partial^{|\beta|} f}{\partial x_1^{\beta_1} \cdots \partial x_n^{\beta_n}}$.

Proposition 3.3.2.3 *For a sequence of functions $H_n(\tilde{x})$ satisfying $H_0(\tilde{x}) = 1$ and $H_{n+1}(\tilde{x}) = \frac{1}{\sqrt{2}}\left(2\tilde{x}H_n(\tilde{x}) - \frac{dH_n(\tilde{x})}{d\tilde{x}}\right)$ we have*

$$\psi_n(\tilde{x}) = H_n(\tilde{x})\psi_0(\tilde{x}).$$

Remark 3.3.2.3 One can check that the family $\{\psi_n\}_{n\in\mathbb{N}}$ forms an orthogonal basis of $L^2(\mathbb{R})$.

Let us now consider the following question: Is the set $\left\{\hbar\omega(n + \frac{1}{2}) \mid n \in \mathbb{N}_0\right\}$ the full spectrum of the operator $\hat{H}$? The answer is yes, but the proof is not straightforward and we will not present it here.

3.3.3 Weyl Quantization on $\mathbb{R}^{2n}$

Let f be a "sufficiently nice" function on $\mathbb{R}^{2n}$, e.g. $f \in \mathcal{S}(\mathbb{R}^{2n})$. We define the *Weyl quantization* (Weyl 1927) $\mathscr{Q}_{\text{Weyl}}(f)$ as an operator on $L^2(\mathbb{R}^n)$ by

$$\mathscr{Q}_{\text{Weyl}}(f) := \frac{1}{(2\pi)^n} \int_{\mathbb{R}^{2n}} \hat{f}(a, b) \underbrace{e^{i(a\hat{x}+b\hat{p})}}_{U(a,b)} da\,db,$$

where $\hat{f}$ denotes the *Fourier transform* of f. We can compute $U(a, b)$ by using the *Baker–Campbell–Hausdorff* (BCH) formula: $e^{A+B} = e^{\frac{1}{2}[A,B]}e^A e^B$ if $[[A, B], B] = [A, [A, B]]$ for any two operators A and B. Formally, we get that

$$U(a, b) = e^{-\frac{1}{2}[ia\hat{x}, ib\hat{p}]}e^{ia\hat{x}}e^{ib\hat{p}} = e^{\frac{i\hbar}{2}ab}e^{ia\hat{x}}e^{ib\hat{p}}.$$

Exercise 3.3.3.1 Show that $\left(e^{ib\hat{p}}\psi\right)(x) = \psi(x + \hbar b)$ for all $\psi \in L^2(\mathbb{R}^n)$.

Using Exercise 3.3.3.1, we get that[4] $U(a, b)\psi(x) = e^{\frac{i\hbar}{2}ab}e^{ia\hat{x}}\psi(x + \hbar b)$.

Moreover, there are some nice properties for the Weyl quantization. In fact, we have that:

- if $f \in \mathcal{S}(\mathbb{R}^{2n})$, then $\mathcal{Q}_{\text{Weyl}}(f)$ is a bounded operator on $L^2(\mathbb{R}^n)$. Actually, it is a *Hilbert–Schmidt* operator.[5]
- the map $\mathcal{Q}_{\text{Weyl}}\colon \mathcal{S}(\mathbb{R}^{2n}) \to L^2(\mathbb{R}^n)$ is a bijection.
- for all $f, g \in \mathcal{S}(\mathbb{R}^{2n})$, we get that $[\mathcal{Q}_{\text{Weyl}}(f), \mathcal{Q}_{\text{Weyl}}(g)] = i\hbar\mathcal{Q}_{\text{Weyl}}(\{f, g\}) + O(\hbar^2)$.

3.4 Schrödinger Equations, Fourier Transforms and Propagators

Recall that in the Hamiltonian formalism of classical mechanics the dynamics (i.e. the time-evolution) were generated by Hamilton's equations associated with a Hamiltonian function on the classical phase space, i.e. by a function $H \in C^\infty(T^*M)$. In quantum mechanics, it is postulated that time-evolution is described by the Schrödinger equation associated with the corresponding Hamiltonian operator $\hat{H}$. In particular, for a Hilbert space $\mathcal{H}$ and some $\psi \in \mathcal{H}$, we consider the initial value problem

$$\begin{cases} i\hbar\frac{\mathrm{d}}{\mathrm{d}t}\psi(t) &= \hat{H}\psi(t), \\ \psi(0) &= \psi_0. \end{cases} \tag{3.4.1}$$

Before we discuss how to solve the Schrödinger equation (SE), let us briefly mention some features of it.

(1) The SE is a *linear* equation, i.e. if $\psi_1(t)$ and $\psi_2(t)$ solve the SE with $\psi_1(0) = \psi_1$ and $\psi_2(0) = \psi_2$, then also $\alpha\psi_1(t) + \beta\psi_2(t)$ solves the SE with initial value

$$\alpha\psi_1(0) + \beta\psi_2(0) = \alpha\psi_1 + \beta\psi_2.$$

Remark 3.4.0.1 The linear SE can easily be generalized to a nonlinear equation but we do not discuss that here.

(2) The SE is *deterministic* in the sense that given $\psi \in \mathcal{H}$, there is a canonical way to produce $\psi(t)$ (we will later formulate this more precisely).
(3) The SE is unitary, i.e. $\|\psi(t)\|_{\mathcal{H}}^2 = \|\psi_0\|_{\mathcal{H}}^2$ for all t. This can be compared with the conservation of energy in classical mechanics.

---

[4] Note that we have used $[\hat{x}, \hat{p}] = i\hbar I$, where I is the identity operator on $\mathcal{H}$.

[5] A Hilbert–Schmidt operator on a Hilbert space $\mathcal{H}$ is a bounded operator $A\colon \mathcal{H} \to \mathcal{H}$ such that $\|A\|_{\text{HS}}^2 := \sum_{i \in I} \|Ae_i\|_{\mathcal{H}}^2 < \infty$, where $\{e_i\}_{i \in I}$ denotes an orthonormal basis of $\mathcal{H}$.

3.4.1 Solving the Schrödinger Equation

We start with a simple situation, namely we assume that $\{\lambda_j\}_{j\in I}$ are eigenvalues of $\hat{H}$ and that the family $\{\phi_{\lambda_j}\}_{j\in I}$ forms an orthonormal basis of the Hilbert space $\mathcal{H}$, where ϕ_{λ_j} denotes an eigenvector associated with the eigenvalue λ_j, i.e. the equation $\hat{H}\phi_{\lambda_j} = \lambda_j\phi_{\lambda_j}$ holds. We want to understand how we can solve the following initial value problem:

$$\begin{cases} i\hbar\dot{\phi}_{\lambda_j}(t) &= \hat{H}\phi_{\lambda_j}(t), \\ \phi_{\lambda_j}(0) &= \phi_{\lambda_j}. \end{cases} \tag{3.4.2}$$

In order to do this, let us first look for solutions of the form

$$\phi_{\lambda_j}(t) = f(t)\phi_{\lambda_j},$$

for some function $f \in C^\infty(\mathbb{R})$. From (3.4.2) it follows that

$$\begin{cases} i\hbar\dot{f}(t)\phi_{\lambda_j} &= \lambda_j f(t)\phi_{\lambda_j}, \\ f(0) &= 1. \end{cases} \tag{3.4.3}$$

Clearly, we can take $f(t) = e^{-\frac{i}{\hbar}t\lambda_j}$, and we immediately see that $\phi_{\lambda_j}(t) = e^{-\frac{i}{\hbar}t\lambda_j}\phi_{\lambda_j}$ solves (3.4.2). Note that we can write

$$\phi_{\lambda_j}(t) = e^{-\frac{i}{\hbar}t\hat{H}}\phi_{\lambda_j}. \tag{3.4.4}$$

Now Eq. (3.4.4) together with the linearity of the SE suggests that "formally" the function

$$\psi(t) = e^{-\frac{i}{\hbar}t\hat{H}}\psi_0 \tag{3.4.5}$$

solves the SE (3.4.1) for all initial values $\psi_0 \in \mathcal{H}$. In fact, if $\psi_0 \in D(\hat{H})$, it can be deduced that $\psi(t) \in D(\hat{H})$ for all t by Stone's theorem, and in this case $\psi(t)$ defined as in (3.4.5) indeed solves the SE (3.4.1). Hence, the function (3.4.5) can be interpreted as a canonical time-evolution of the initial value $\psi_0 \in \mathcal{H}$. This is what is usually referred to as the *deterministic* feature of the SE.

Remark 3.4.1.1 In order to define $\psi(t) = e^{-\frac{i}{\hbar}t\hat{H}}\psi_0$ we do not need the assumption that it has an eigenbasis, we only need the operator $\hat{H}$ to be self-adjoint.

Definition 3.4.1.1 (*Propagator*) The operator $U(t) = e^{-\frac{i}{\hbar}t\hat{H}}$ is usually called (the quantum mechanical) *propagator*.

Lemma 3.4.1.1 *If $\{\phi_{\lambda_j}\}_{j\in I}$ is an eigenbasis with respect to $\hat{H}$ with ϕ_{λ_j} being eigenvectors associated with the eigenvalues λ_j, we get that*

$$U(t) = \sum_{j=1}^{n} e^{-\frac{i}{\hbar}t\lambda_j} \phi_{\lambda_j}^* \otimes \phi_{\lambda_j}, \qquad (3.4.6)$$

where $\phi_{\lambda_j}^* \in \mathcal{H}^*$ is the dual basis element of ϕ_{λ_j}.

Proof Let $\psi \in \mathcal{H}$ and write it as a linear combination $\psi = \sum_{k=1}^{n} c_k \phi_{\lambda_k}$ with $c_k \in \mathbb{C}$ for $1 \leq k \leq n$. We know that

$$U(t)\psi = \sum_{k=1}^{n} c_k U(t)\phi_{\lambda_k} = \sum_{k=1}^{n} c_k e^{-\frac{i}{\hbar}t\lambda_k} \phi_{\lambda_k}. \qquad (3.4.7)$$

On the other hand, we have that

$$\left(\sum_{j=1}^{n} e^{-\frac{i}{\hbar}t\lambda_j} \phi_{\lambda_j}^* \otimes \phi_{\lambda_j}\right)\psi = \sum_{k,j=1}^{n} c_k e^{-\frac{i}{\hbar}t\lambda_k} \phi_{\lambda_j} \underbrace{\phi_{\lambda_j}^*(\phi_{\lambda_k})}_{=\delta_{jk}} = \sum_{k=1}^{n} c_k e^{-\frac{i}{\hbar}t\lambda_k} \phi_{\lambda_k}.$$

$$(3.4.8)$$

Thus, for all ψ, we get

$$U(t)\psi = \left(\sum_{j=1}^{n} e^{-\frac{i}{\hbar}t\lambda_j} \phi_{\lambda_j}^* \otimes \phi_{\lambda_j}\right)\psi.$$

$\square$

Let us give a short summary of the discussion so far.

- The operator $U(t) = e^{-\frac{i}{\hbar}t\hat{H}}$ can be used to describe time-evolution of states in a canonical way.
- If $\hat{H}$ has an eigenbasis $\{\phi_{\lambda_j}\}$, corresponding to the eigenvalues λ_j, then $U(t)$ can be described explicitly as

$$U(t) = \sum_{j=1}^{n} e^{-\frac{i}{\hbar}t\hat{H}} \phi_{\lambda_j}^* \otimes \phi_{\lambda_j}.$$

3.4.2 The Schrödinger Equation for the Free Particle Moving on $\mathbb{R}$

In this situation we have that $\mathcal{H} = L^2(\mathbb{R})$ and $\hat{H} = \frac{1}{2m}\hat{p}^2 = -\frac{\hbar^2}{2m}\frac{d^2}{dx^2}$. Note that there is no interaction (potential) term in the Hamiltonian operator. Hence, the SE (3.4.1) becomes

$$\begin{cases} i\hbar \frac{\partial}{\partial t} \psi(x,t) &= -\frac{\hbar^2}{2m} \frac{\partial^2}{\partial x^2} \psi(x,t), \\ \psi(x,0) &= \psi(x). \end{cases} \qquad (3.4.9)$$

We will now explain how one can solve (3.4.9) with the method of *Fourier transform* and will try to find an explicit representation of the propagator $U(t)$.

3.4.2.1　Digression on Fourier Transform

Let us briefly recall the definition and properties of the Fourier transform. Let $\mathcal{S}(\mathbb{R}^n)$ be the space of Schwartz functions on $\mathbb{R}^n$. Recall that $f \in \mathcal{S}(\mathbb{R}^n)$ roughly means that $f \in C^\infty(\mathbb{R}^n)$ and f and all its derivatives approach zero as $|x| \to \infty$ faster than any polynomial function approaches infinity. Now consider a function $f \in \mathcal{S}(\mathbb{R}^n)$. The *Fourier transform* $\mathcal{F}(f)$, or sometimes simply $\hat{f}$, of f is defined by

$$\hat{f}(k) := \frac{1}{(2\pi)^{\frac{n}{2}}} \int_{\mathbb{R}^n} e^{-i\langle k,x \rangle} f(x) \mathrm{d}x, \qquad (3.4.10)$$

where $\langle \cdot, \cdot \rangle : \mathbb{R}^n \times \mathbb{R}^n \to \mathbb{R}$ denotes the standard inner product on $\mathbb{R}^n$. We list some properties of the Fourier transform without proofs:

(i)　For any function $f \in \mathcal{S}(\mathbb{R}^n)$ we have $\hat{f} \in \mathcal{S}(\mathbb{R}^n)$.
(ii)　For any function $f \in \mathcal{S}(\mathbb{R}^n)$ we get that

$$\widehat{\frac{\partial f}{\partial x_j}} = i k_j \hat{f}, \qquad (3.4.11)$$

$$\widehat{x_j f} = i \frac{\partial \hat{f}}{\partial k_j}. \qquad (3.4.12)$$

(iii)　For any function $f \in \mathcal{S}(\mathbb{R}^n)$ we can define the *inverse Fourier transform*

$$\mathcal{F}^{-1}(\hat{f})(x) := f(x) = \frac{1}{(2\pi)^{\frac{n}{2}}} \int_{\mathbb{R}^n} e^{i\langle k,x \rangle} \hat{f}(k) \mathrm{d}k. \qquad (3.4.13)$$

(iv)　For any function $f \in \mathcal{S}(\mathbb{R}^n)$ we have *Plancherel's formula*

$$\int_{\mathbb{R}^n} |f(x)|^2 \mathrm{d}x = \int_{\mathbb{R}^n} |\hat{f}(k)|^2 \mathrm{d}k. \qquad (3.4.14)$$

(v)　We have the following theorem:

Theorem 3.4.2.1 (Combined inversion and Plancherel's formula) *The Fourier transform* $\mathcal{F}: \mathcal{S}(\mathbb{R}^n) \to \mathcal{S}(\mathbb{R}^n)$ *can be extended to a unique bounded map* $\mathcal{F}: L^2(\mathbb{R}^n) \to L^2(\mathbb{R}^n)$. *This map can be computed as*

$$\mathcal{F}(f)(k) = \frac{1}{(2\pi)^{\frac{n}{2}}} \lim_{A \to \infty} \int_{|x| \le A} e^{-i\langle k, x \rangle} f(x) \mathrm{d}x. \qquad (3.4.15)$$

Moreover, the inverse Fourier transform $\mathcal{F}^{-1} : L^2(\mathbb{R}^n) \to L^2(\mathbb{R}^n)$ *is unitary and*

$$\mathcal{F}^{-1}(f)(k) = \frac{1}{(2\pi)^{\frac{n}{2}}} \lim_{A \to \infty} \int_{|x| \le A} e^{i\langle k, x \rangle} \hat{f}(k) \mathrm{d}k. \qquad (3.4.16)$$

Remark 3.4.2.1 If $f \in L^1(\mathbb{R}^n) \cap L^2(\mathbb{R}^n)$, then

$$\mathcal{F}(f)(k) = \frac{1}{(2\pi)^{\frac{n}{2}}} \int_{\mathbb{R}^n} e^{-i\langle k, x \rangle} f(x) \mathrm{d}x,$$

because in this case

$$\lim_{A \to \infty} \int_{|x| \le A} e^{-i\langle k, x \rangle} f(x) \mathrm{d}x = \int_{\mathbb{R}^n} e^{-i\langle k, x \rangle} f(x) \mathrm{d}x$$

by Lebesgue's dominated convergence theorem.

(vi) Let f and g be two measurable functions on $\mathbb{R}^n$. The *convolution* product of f and g is defined as

$$(f * g)(x) := \int_{\mathbb{R}^n} f(x - y) g(y) \mathrm{d}y,$$

where we assume that the right-hand side exists. Moreover, if $f, g \in L^1(\mathbb{R}^n) \cap L^2(\mathbb{R}^n)$, then we get that

$$\frac{1}{(2\pi)^{\frac{n}{2}}} \mathcal{F}(f * g) = \mathcal{F}(f)\mathcal{F}(g).$$

3.4.3 Solving the Schrödinger Equation with Fourier Transform

Let us first look for solutions of the form $\psi(x, t) = e^{i(kx - \omega(k)t)}$. From (3.4.9) it is clear that $\psi(x, t)$ is a solution if and only if $\omega(k) = \frac{\hbar k^2}{2m}$. Hence, we get that

$$\psi(x, t) = e^{ikx - i\frac{\hbar k^2}{2m}t} \qquad (3.4.17)$$

is a solution of the SE. However, note that such $\psi(x, t) \notin L^2(\mathbb{R}^n)$. Therefore, $\psi(x, t)$ is not the solution we are looking for. Here, the idea is to use $\psi(x, t)$ to produce a sensible solution of (3.4.9).

Proposition 3.4.3.1 *Let* $\psi_0 \in \mathcal{S}(\mathbb{R})$ *and let* $\hat{\psi}_0$ *be its Fourier transform. Define the function*

$$\psi(x, t) := \frac{1}{(2\pi)^{\frac{1}{2}}} \int_{\mathbb{R}} \hat{\psi}_0(k) e^{i(kx - \omega(k)t)} dk. \tag{3.4.18}$$

Then the function $\psi(x, t)$ *is a solution of (3.4.9) with* $\psi(x, 0) = \psi_0(x)$.

Proof Since $\hat{\psi}_0(k) \in \mathcal{S}(\mathbb{R})$, we can check that the derivatives with respect to x and t can be interchanged with the integral sign in the definition of $\psi(x, t)$. Since $e^{i(kx - \omega(k)t)}$ solves the SE, we can easily check that $\psi(x, t)$ solves (3.4.9). Moreover, we get that

$$\psi(x, 0) = \frac{1}{(2\pi)^{\frac{1}{2}}} \int_{\mathbb{R}} e^{ikx} \hat{\psi}_0(k) dt = \psi_0(x),$$

where the last equality holds because of the inverse Fourier transform.   □

Corollary 3.4.3.1 *Let* ψ_0 *be as in Proposition 3.4.3.1 and let* $\hat{\psi}(k, t)$ *be the Fourier transform of* $\psi(x, t)$ *with respect to* t. *Then we get that*

$$\hat{\psi}(x, t) = \hat{\psi}_0(k) e^{-i\omega(k)t}.$$

Proof From Proposition 3.4.3.1 we know that

$$\psi(x, t) = \frac{1}{(2\pi)^{\frac{1}{2}}} \int_{\mathbb{R}} e^{ikx} \left(e^{i\omega(k)t} \hat{\psi}_0(k) \right) dk.$$

Thus, the claim follows.   □

From property *(vi)* of Fourier transforms, formally we get

$$e^{-i\omega(k)t} \hat{\psi}_0(k) = \frac{1}{(2\pi)^{\frac{1}{2}}} \mathcal{F}(K_t * \psi_0),$$

where $\mathcal{F}(K_t) = e^{-i\omega(k)t}$, which implies that

$$K_t = \mathcal{F}^{-1} \left(e^{-i\omega(k)t} \right) = \frac{1}{(2\pi)^{\frac{1}{2}}} \int_{\mathbb{R}} e^{ikx} e^{-i\omega(k)t} dk.$$

Again, a "formal computation" shows that

$$K_t(x) = \sqrt{\frac{m}{i2\pi\hbar t}} e^{\frac{imx^2}{2t\hbar}}.$$

The computation of $K_t(x)$ is "formal" because $e^{-i\omega(k)t} \notin L^1(\mathbb{R}) \cap L^2(\mathbb{R})$ and thus we do not know how to take the inverse Fourier transform of it. Hence, we need a way to make sense of an integral of the form

$$\int_{\mathbb{R}} e^{ikx} e^{-i\omega(k)t} \, dk. \tag{3.4.19}$$

Integrals of the form (3.4.19) are called *Fresnel integrals*.

3.4.3.1 Digression on Fresnel Integrals

We want to look more closely to the properties of Fresnel integrals. Let Q be a real, symmetric $n \times n$ matrix with $\det(Q) \neq 0$. An integral of the form

$$\int_{\mathbb{R}^n} e^{\frac{i}{2}\langle Qx, x \rangle} \, dx$$

is called a *Fresnel integral*, and is defined as

$$\int_{\mathbb{R}^n} e^{\frac{i}{2}\langle Qx, x \rangle} \, dx := \lim_{\varepsilon \to 0} \int_{\mathbb{R}^n} e^{-\frac{1}{2}\varepsilon\langle x, x \rangle} e^{\frac{i}{2}\langle Qx, x \rangle} \, dx.$$

As a matter of fact we have

$$\int_{\mathbb{R}^n} e^{\frac{i}{2}\langle Qx, x \rangle} \, dx = e^{\frac{\pi i}{4} \operatorname{sign}(Q)} \frac{1}{\left| \det\left(\frac{Q}{2\pi} \right) \right|^{\frac{1}{2}}},$$

where $\operatorname{sign}(Q) = \#$positive eigenvalues $- \#$negative eigenvalues. More generally, for $\omega \in \mathbb{R}^n$, we have

$$\int_{\mathbb{R}^n} e^{\frac{i}{2}\langle Qx, x \rangle} e^{\langle \omega, x \rangle} \, dx = \lim_{\varepsilon \to 0} \int_{\mathbb{R}^n} e^{\frac{i}{2}\langle Qx, x \rangle - \frac{1}{2}\varepsilon\langle x, x \rangle} e^{\langle \omega, x \rangle} \, dx$$

$$= \frac{e^{\frac{\pi i}{4} \operatorname{sign}(Q)}}{\left| \det\left(\frac{Q}{2\pi} \right) \right|^{\frac{1}{2}}} e^{\frac{i}{2}\langle Q^{-1}\omega, \omega \rangle}. \tag{3.4.20}$$

We use this general result to compute the integral

$$\frac{1}{(2\pi)^{\frac{1}{2}}} \int_{\mathbb{R}} e^{-i\omega(k)t} e^{ikx} \, dk. \tag{3.4.21}$$

Now, using (3.4.20), it can be easily checked that

$$K_t(x) = \sqrt{\frac{m}{2\pi i k t}} e^{\frac{imx^2}{2t\hbar}}.$$

Remark 3.4.3.1 There is also another way to define (3.4.21) (see Hall (2013)).

Let us now consider a rigorous approach to the previous formal discussion within the following proposition.

Proposition 3.4.3.2 *Suppose $\psi_0 \in L^1(\mathbb{R}) \cap L^2(\mathbb{R})$ and define*

$$\psi(x, t) := \mathcal{F}^{-1}\left(\hat{\psi}_0(k)e^{-\frac{\hbar k^2 t}{2m}}\right).$$

Then $\psi(x, t) = K_t * \psi_0$, where $K_t(x) = \sqrt{\frac{m}{2\pi i k t}}e^{\frac{imx^2}{2t\hbar}}$.

Proof We will only briefly sketch the proof. The idea here is to show that

$$\mathcal{F}(K_t * \psi_0) = \hat{\psi}_0(k)e^{-\frac{i\hbar k^2 t}{2m}}. \tag{3.4.22}$$

Since $K_t \notin L^2(\mathbb{R})$, we cannot actually talk about its Fourier transform $\mathcal{F}(K_t)$. However, we can instead consider $K_t \chi_{[-n,n]}$ and its Fourier transform. We can observe that

$$\frac{1}{(2\pi)^{\frac{1}{2}}}\mathcal{F}(K_t \chi_{[-n,n]} * \psi_0) = \mathcal{F}(K_t \chi_{[-n,n]})\mathcal{F}(\psi_0).$$

It can be shown that $K_t \chi_{[-n,n]} * \psi_0 \xrightarrow{n \to \infty} K_t * \psi$ in $L^2(\mathbb{R})$ and

$$\mathcal{F}(K_t \chi_{[-n,n]})\mathcal{F}(\psi_0) \xrightarrow{n \to \infty} \frac{1}{(2\pi)^{\frac{1}{2}}}e^{-\frac{i\hbar k^2 t}{2m}}\hat{\psi}_0$$

in $L^2(\mathbb{R})$. These two observations imply that (3.4.22) holds and hence we get that

$$K_t * \psi_0 = \mathcal{F}^{-1}\left(\hat{\psi}_0(k)e^{-\frac{\hbar k^2 t}{2m}}\right).$$

$\square$

3.4.3.2 Summary of the Discussion

We have shown that if $\psi_0 \in L^1(\mathbb{R}) \cap L^2(\mathbb{R})$, then

$$e^{-\frac{i}{\hbar}t\hat{H}}\psi_0 = \left(\mathcal{F}^{-1} \circ \mathsf{m} \circ \mathcal{F}\right)\psi_0,$$

where $(\mathsf{m}f)(k) := e^{-\frac{i\hbar k^2 t}{2m}}f(k)$. This means we have shown that the following diagram is commutative:

$$
\begin{array}{ccc}
L^1(\mathbb{R}) \cap L^2(\mathbb{R}) & \xrightarrow{\ \mathcal{F}\ } & L^1(\mathbb{R}) \cap L^2(\mathbb{R}) \\
{\scriptstyle e^{-\frac{i}{\hbar}t\hat{H}}}\Big\downarrow & & \Big\downarrow{\scriptstyle \mathsf{m}} \\
L^2(\mathbb{R}) & \xleftarrow[\ \mathcal{F}^{-1}\ ]{} & L^2(\mathbb{R})
\end{array}
$$

Moreover, we have shown that $(\mathcal{F}^{-1} \circ \mathsf{m} \circ \mathcal{F})\psi_0 = K_t * \psi_0$. Finally, combining these results, we conclude that

$$\left(e^{-\frac{i}{\hbar}t\hat{H}}\psi_0\right)(x) = \sqrt{\frac{m}{2\pi i\hbar t}} \int_{\mathbb{R}} e^{\frac{im(x-y)^2}{2t\hbar}} \psi_0(y)dy,$$

i.e. the integral kernel of $e^{-\frac{i}{\hbar}t\hat{H}}$ is $K_t(x-y) = \sqrt{\frac{m}{2\pi i\hbar t}}e^{\frac{im(x-y)^2}{2t\hbar}}$.

Remark 3.4.3.2 One can check that $K_t(x)$ satisfies the SE and $\lim_{t\to 0} K_t(x) = \delta(x)$ in the *distributional* sense.

Definition 3.4.3.1 (*Fundamental solution*) $K_t(x)$ is called the *fundamental solution* of the SE.

Remark 3.4.3.3 One can easily extend the discussion above to the free particle in $\mathbb{R}^n$.

Chapter 4
The Path Integral Approach to Quantum Mechanics

We have seen that the Hamiltonian approach to classical mechanics can be used to consider an axiomatic approach to quantum mechanics. Hence, it is natural to ask whether there is a "Lagrangian formulation" of quantum mechanics, i.e., a formulation where the classical action functional $S[\gamma]$ is used. Dirac, who considered Lagrangian mechanics as a more fundamental approach to a classical system, took the first steps towards a Lagrangian formulation of quantum mechanics (Dirac 1930, 1933). Dirac's approach was extended and formalized by Feynman, who gave a formulation through the mathematical concept of functional integrals (Feynman 1942; Feynman and Hibbs 1965; Zinn Justin 2004), also called a *Feynman path integral* formulation, for a quantum theory. Moreover, Dirac suggested that the propagator $K(t, x, y)$ in quantum mechanics can be represented by an object of the form

$$\int_{\gamma \in P(\mathbb{R}, x, y)} e^{\frac{i}{\hbar} S[\gamma]} \mathscr{D}[\gamma], \tag{4.0.1}$$

where $P(\mathbb{R}, x, y)$ is the space of paths $\gamma \colon [0, t] \to \mathbb{R}$ joining two positions x and y, and $\mathscr{D}$ denotes a "symbol" for a measure on $P(\mathbb{R}, x, y)$. Since $P(\mathbb{R}, x, y)$ is an infinite-dimensional manifold, it is not clear what the actual meaning of the integral (4.0.1) measure-theoretically is because no measure exists on this space. Thus, we need to figure out how one can still make sense of an object of the form (4.0.1).

4.1 Feynman's Formulation of the Path Integral

Feynman's idea (Feynman 1942; Feynman and Hibbs 1965) was to define (4.0.1) as a limit of integrals over finite-dimensional manifolds, which roughly works as follows. Let $P_n(\mathbb{R}, x, y)$ be the space of piece-wise linear paths joining x to y, which consists of n line segments $\ell_{x,x_1}, \ell_{x_1,x_2}, \ldots, \ell_{x_{n-1},y}$. Clearly, to define $\gamma \in P_n(\mathbb{R}, x, y)$, we need to specify the tuple $(x_1, \ldots, x_{n-1})$. This means that we can identify $P_n(\mathbb{R}, x, y)$ with

© The Author(s), under exclusive license to Springer Nature Singapore Pte Ltd. 2023
N. Moshayedi, *Quantum Field Theory and Functional Integrals*,
SpringerBriefs in Physics, https://doi.org/10.1007/978-981-99-3530-7_4

$\mathbb{R}^{n-1}$. Hence, we can define

$$\int_{\gamma \in P(\mathbb{R}, x, y)} e^{\frac{i}{\hbar} S[\gamma]} \mathscr{D}[\gamma] := \lim_{n \to \infty} A(n, t) \int_{\gamma \in P_n(\mathbb{R}, x, y)} e^{\frac{i}{\hbar} S[\gamma]} dx_1 \cdots dx_{n-1}, \quad (4.1.1)$$

where $A(n, t)$ is some constant depending on n and t.

4.1.1   The Free Propagator for the Free Particle on $\mathbb{R}$

We have already shown that the propagator is given by

$$K(t, x, y) = \sqrt{\frac{m}{2\pi i \hbar t}} e^{\frac{i}{\hbar} \frac{m}{2t} (x-y)^2}. \tag{4.1.2}$$

Let us now give a path integral derivation of $K(t, x, y)$. Let $0 = t_0 < \cdots < t_n = t$ with $t_i - t_{i-1} = \frac{t}{n} =: \Delta t$. Moreover, let $(x_1, ..., x_{n-1}) \in \mathbb{R}^{n-1}$ and let γ be the piecewise linear path joining x to y such that $\gamma(t_i) = x_i$ and the line segment joining x_{i-1} to x_i is given by

$$\gamma(s) = \frac{1}{\Delta t} \left((t_i - s) x_{i-1} + (s - t_{i-1}) x_i \right), \quad s \in [t_{i-1}, t_i], \quad i = 1, 2, ..., n$$

Then we can write the action as

$$S[\gamma] = \frac{1}{2} m \sum_{i=1}^{n} \int_{t_{i-1}}^{t_i} \frac{(x_i - x_{i-1})^2}{(\Delta t)^2} ds = \frac{1}{2} m \sum_{i=1}^{n} \frac{(x_i - x_{i-1})^2}{\Delta t} \tag{4.1.3}$$

and thus we get the expression

$$A(n, t) \int_{\mathbb{R}^{n-1}} e^{\frac{i}{\hbar} S[\gamma]} dx_1 \cdots dx_{n-1} = A(n, t) \int_{\mathbb{R}^{n-1}} e^{\frac{i}{\hbar} \frac{m}{2} \sum_{i=1}^{n} \frac{(x_i - x_{i-1})^2}{\Delta t}} dx_1 \cdots dx_{n-1}.$$

$$(4.1.4)$$

Let us now define $f_i := \sqrt{\frac{m}{2\hbar \Delta t}} x_i$. Then, by change of variables, this integral will be

$$A(n, t) \left(\frac{2\hbar \Delta t}{m} \right)^{\frac{n-1}{2}} \int_{\mathbb{R}^{n-1}} e^{i \sum_{i=1}^{n} (f_i - f_{i-1})^2} df_1 \cdots df_{n-1} \tag{4.1.5}$$

$$= A(n, t) \left(\frac{2\hbar \Delta t}{m} \right)^{\frac{n-1}{2}} \frac{(\pi i)^{\frac{n-1}{2}}}{\sqrt{n}} e^{\frac{i}{\hbar} (f_n - f_1)^2} = A(n, t) \left(\frac{2\hbar \Delta t}{m} \right)^{\frac{n-1}{2}} \frac{(\pi i)^{\frac{n-1}{2}}}{\sqrt{n}} e^{\frac{i}{\hbar} \frac{m}{2n\Delta t} (x_n - x_1)^2}$$

$$= A(n, t) \left(\frac{2\pi i \hbar \Delta t}{m} \right)^{\frac{n-1}{2}} \left(\frac{m}{2n\pi i \hbar \Delta t} \right)^{\frac{1}{2}} e^{\frac{i}{\hbar} \frac{m}{2n\Delta t} (x_n - x_1)^2}$$

$$= A(n, t) \left(\frac{2\pi i \hbar \Delta t}{m} \right)^{\frac{n-1}{2}} \left(\frac{m}{2n\pi i \hbar \Delta t} \right)^{\frac{1}{2}} e^{\frac{i}{\hbar} \frac{m}{2t} (y-x)^2}.$$

Let us now define the constant to be $A(n, t) := \left(\frac{m}{2\pi i \hbar t}\right)^{\frac{n}{2}}$. Then we get that

$$\int_{\gamma \in P(\mathbb{R}, x, y)} e^{\frac{i}{\hbar} S[\gamma]} \mathscr{D}[\gamma] = \lim_{n \to \infty} A(n, t) \int_{\mathbb{R}^{n-1}} e^{\frac{i}{\hbar} S[\tilde{\gamma}]} dx_1 \cdots dx_{n-1}$$

$$= \left(\frac{m}{2\pi i \hbar t}\right)^{\frac{1}{2}} e^{\frac{i}{\hbar} \frac{m}{2t}(x-y)^2} \qquad (4.1.6)$$

$$= K(t, x, y).$$

Let us now see how we can derive the path integral representation of the propagator associated with a Hamiltonian of the form $\hat{H} = \hat{H}_0 + V(\hat{x})$, where $\hat{H}_0 = \frac{1}{2m} \hat{p}^2$ is the *free Hamiltonian*. We first want to recall the *Kato–Lie–Trotter (KLT) product formula*. Let A and B be self-adjoint operators on a Hilbert space $\mathcal{H}$ with domains $D(A)$ and $D(B)$, respectively. Assume that $A + B$ is densely defined and essentially self-adjoint on $D(A) \cap D(B)$. Then the KLT product formula reads

$$\lim_{n \to \infty} \left(e^{\frac{i}{n} t A} e^{\frac{i}{n} t B}\right)^n = e^{it(A+B)} \qquad (4.1.7)$$

in the strong operator topology, i.e., $A_n \to A$ if and only if $\|A_n \psi - A\psi\| \xrightarrow{n \to \infty} 0$ for all $\psi \in \mathcal{H}$. Now assume that $V(\hat{x})$ is "sufficiently nice" such that the assumption of the KLT product formula is satisfied. Then for all $\psi \in L^2(\mathbb{R})$, we have

$$e^{-\frac{i}{\hbar} t(\hat{H}_0 + V(\hat{x}))} \psi = \lim_{n \to \infty} \left(e^{-\frac{i}{\hbar} \frac{t}{n} \hat{H}_0} e^{-\frac{i}{\hbar} \frac{t}{n} V(\hat{x})}\right)^n \psi. \qquad (4.1.8)$$

Let us first compute the right-hand side of (4.1.8). Recall that

$$\left(e^{-\frac{i}{\hbar} \frac{t}{n} \hat{H}_0} \psi\right)(x_1) = \sqrt{\frac{m}{2\pi i \hbar \frac{t}{n}}} \int_{\mathbb{R}} e^{\frac{i}{\hbar} \frac{m}{2\frac{t}{n}}(x_1 - x_0)^2} dx_0 \qquad (4.1.9)$$

and thus we have that

$$\left(e^{-\frac{i}{\hbar} \frac{t}{n} V(\hat{x})} \psi\right)(x) = e^{-\frac{i}{\hbar} \frac{t}{n} V(x)} \psi(x). \qquad (4.1.10)$$

Using these two relations, we get that

$$\left(\left(e^{-\frac{i}{\hbar} \frac{t}{n} \hat{H}_0} e^{-\frac{i}{\hbar} \frac{t}{n} V(\hat{x})}\right) \psi\right)(x_1) = \sqrt{\frac{m}{2\pi i \hbar \frac{t}{n}}} \int_{\mathbb{R}} e^{\frac{i}{\hbar} \frac{m}{2\frac{t}{n}}(x_1 - x_0)^2} e^{-\frac{i}{\hbar} \frac{t}{n} V(x_0)} \psi(x_0) dx_0.$$

$$(4.1.11)$$

Repeating this procedure, we get that

$$\left(\left(e^{-\frac{i}{\hbar} \frac{t}{n} \hat{H}_0} e^{-\frac{i}{\hbar} \frac{t}{n} V(\hat{x})}\right)^n \psi\right)(x_n)$$

$$= \left(\frac{m}{2\pi i\hbar \frac{t}{n}}\right)^{\frac{n}{2}} \int_{\mathbb{R}^n} e^{\frac{i}{\hbar}\frac{m}{2}\frac{t}{n}\sum_{k=1}^{n}(x_k - x_{k-1})^2 - \frac{i}{\hbar}\frac{t}{n}\sum_{k=1}^{n} V(x_{k-1})} \psi(x_0)dx_0 dx_1 \cdots dx_{n-1}$$

$$= \left(\frac{m}{2\pi i\hbar \frac{t}{n}}\right)^{\frac{n}{2}} \int_{\mathbb{R}^n} e^{\frac{i}{\hbar}\sum_{k=1}^{n}\frac{t}{n}\left\{\frac{m}{2}\left(\frac{x_k - x_{k-1}}{\frac{t}{n}}\right)^2 - V(x_{k-1})\right\}} \psi(x)dx dx_1 \cdots dx_{n-1}$$

$$= \int_{\mathbb{R}} \left\{ \left(\frac{m}{2\pi i\hbar \frac{t}{n}}\right)^{\frac{n}{2}} \int_{\mathbb{R}^{n-1}} e^{\frac{i}{\hbar}\sum_{k=1}^{n}\frac{t}{n}\left\{\frac{m}{2}\left(\frac{x_k - x_{k-1}}{\frac{t}{n}}\right)^2 - V(x_{k-1})\right\}} dx_1 \cdots dx_{n-1} \right\} \psi(x_0)dx_0.$$

Taking the limit for $n \to \infty$, we get

$$\lim_{n\to\infty}\left(\left(e^{-\frac{i}{\hbar}\frac{t}{n}\hat{H}_0}e^{-\frac{i}{\hbar}\frac{t}{n}V(\hat{x})}\right)^n \psi\right)(x_n) = \tag{4.1.12}$$

$$= \int_{\mathbb{R}} \left\{ \lim_{n\to\infty}\left(\frac{m}{2\pi i\hbar \frac{t}{n}}\right)^{\frac{n}{2}} \int_{\mathbb{R}^{n-1}} e^{\frac{i}{\hbar}\sum_{k=1}^{n}\frac{t}{n}\left\{\frac{m}{2}\left(\frac{x_k - x_{k-1}}{\frac{t}{n}}\right)^2 - V(x_{k-1})\right\}} dx_1 \cdots dx_{n-1} \right\} \psi(x_0)dx_0$$

$$= \int_{\mathbb{R}} K(t, x, x_0)\psi(x_0)dx_0,$$

where

$$K(t, x, x_0) = \lim_{n\to\infty}\left(\frac{m}{2\pi i\hbar \frac{t}{n}}\right)^{\frac{n}{2}} \int_{\mathbb{R}^{n-1}} e^{\frac{i}{\hbar}\sum_{k=1}^{n}\frac{t}{n}\left\{\frac{m}{2}\left(\frac{x_k - x_{k-1}}{\frac{t}{n}}\right)^2 - V(x_{k-1})\right\}} dx_1 \cdots dx_{n-1}.$$

Moreover, we observe that

$$\lim_{n\to\infty}\sum_{k=1}^{n}\frac{t}{n}\left\{\frac{m}{2}\left(\frac{x_k - x_{k-1}}{\frac{t}{n}}\right)^2 - V(x_{k-1})\right\}$$

can actually be interpreted as

$$\int_0^t \left(\frac{m}{2}\|\dot{\gamma}(s)\|^2 - V(\gamma(s))\right)ds.$$

This means we can finally deduce that

$$\int_{\gamma \in P(\mathbb{R}, x, y)} e^{\frac{i}{\hbar}S[\gamma]}\mathscr{D}[\gamma] := \lim_{n\to\infty}\left(\frac{m}{2\pi i\hbar \frac{t}{n}}\right)^{\frac{1}{2}} \int_{\mathbb{R}^{n-1}} e^{\frac{i}{\hbar}S[\gamma]}dx_1 \cdots dx_{n-1} = K(t, x, y).$$

$$\tag{4.1.13}$$

4.2 Construction of the Wiener Measure

What we have seen so far is that Feynman defined the path integral $\int_{\gamma \in P(\mathbb{R}, x, y)}$ $e^{\frac{i}{\hbar} S[\gamma]} \mathscr{D}[\gamma]$ as a limit of integrals over finite-dimensional manifolds. As a next step, we would like to understand whether it is possible to define a probability measure on $P(\mathbb{R}, x, y)$, which is of the form

$$\frac{e^{\frac{i}{\hbar} S[\gamma]} \mathscr{D}[\gamma]}{Z},$$

where Z is some quantity responsible for the *normalization* of the measure, i.e., to guarantee that the integral over the whole space is equal to one. A short answer to this question is that it is *not* possible However, if we replace i by -1, i.e., we consider a so-called *Wick rotation*, it will be actually possible to construct a measure of the desired form on a suitable $P(\mathbb{R}, x, y)$. This construction was done by *Wiener* in 1923 for the case where $V(x) = 0$ and the resulting measure is known today as the *Wiener measure* (Wiener 1923). From now on we assume for simplicity that $V(x) = 0$ and $S[\gamma] = \frac{1}{2} \int_0^t \|\dot{\gamma}(s)\|^2 ds$. The main ideas of the construction are the following:

- We can interpret

$$A(n, t) \int_{\substack{E \subseteq \mathbb{R}^{n-1} \\ \text{measurable}}} e^{-S[\gamma]} dx_1 \cdots dx_{n-1} \qquad (4.2.1)$$

  as a measure of a certain "measurable" subset E of $P(\mathbb{R}, x, y)$.
- Instead of taking the limit $n \to \infty$, we try to extend the "measure", defined by (4.2.1), to a measure on $P(\mathbb{R}, x, y)$.

The essential idea comes from the kinetic theory of molecules in thermodynamics. It was Einstein (Einstein 1908) who showed that, if $\rho(x, t)$ is the probability density for finding the *Brownian particle* at location x and at time t, then it satisfies the *diffusion equation*

$$\frac{\partial}{\partial t} \rho(x, t) = D \frac{\partial^2}{\partial x^2} \rho(x, t), \qquad (4.2.2)$$

where D is the *diffusion constant*. This immediately implies that

$$\rho(x, t) = \frac{1}{\sqrt{4\pi D t}} e^{\frac{x^2}{4Dt}},$$

if we insist that $\lim_{t \to 0} \rho(x, t) = \delta_0(x)$, where δ_0 is the *Dirac delta function*. This implies that for any measurable set $E \subseteq \mathbb{R}$, the probability of finding the Brownian particle in E at time t is given by

$$\frac{1}{\sqrt{4\pi D t}} \int_E e^{-\frac{x^2}{4Dt}} dx. \qquad (4.2.3)$$

In our case we want to consider $D = \frac{1}{2}$. More generally, we can see that $\frac{1}{\sqrt{2\pi(t_2-t_1)}} e^{\frac{(x-y)^2}{2(t_2-t_1)}}$ is the probability density of finding the particle at position y at time $t = t_2$ if it was at position x at time $t = t_1$. This means, given time steps $0 = t_0 < t_1 < \cdots < t_n \leq t$ and the measurable set $E = \prod_{i=1}^{n} (\alpha_i, \beta_i]$, with $\alpha_i, \beta_i \in \mathbb{R}$ with $\alpha_i \leq \beta_i$ for all $i = 1, \ldots, n$, we can observe that

$$A(n, t) \int_E e^{-\frac{1}{2} \sum_{i=1}^{n} \frac{(x_i - x_{i-1})^2}{t_i - t_{i-1}}} \, dx_1 \cdots dx_n \tag{4.2.4}$$

can be interpreted as the probability of finding the Brownian particle in the interval $(\alpha_i, \beta_i]$ at time $t = t_i$. Hence, it should not be surprising to interpret (4.2.1) as a "measure" of a suitable subset of $P(\mathbb{R}, x, y)$. We want to make this precise and construct the Wiener measure. First, we need some notations and definitions.

- We will write

$$C_0([0, 1]) := \{x \colon [0, 1] \to \mathbb{R} \mid x \text{ is continuous at } x(0)\},$$

which are paths starting at 0. Recall that $C_0([0, 1])$ is actually a Banach space with norm given by

$$\|x\| = \sup_{t \in [0,1]} |x(t)|.$$

Hence, it is a topological space. Let $\mathcal{B}(C_0([0, 1]))$ denote the Borel sigma algebra of $C_0([0, 1])$ with respect to the topology induced by the norm $\| \cdot \|$.

- Fix some $t \in [0, 1]$ and define the map $\mathrm{ev}_t \colon C_0([0, 1]) \to \mathbb{R}$, $\mathrm{ev}_t(x) = x(t)$. It is then not hard to see that ev_t is continuous and hence it is Borel measurable. More generally, given $t_1, \ldots, t_n \in [0, 1]$, we can define the map

$$P(t_1, \ldots, t_n) \colon C_0([0, 1]) \longrightarrow \mathbb{R}^n,$$
$$x \longmapsto P(t_1, \ldots, t_n)(x) = (x(t_1), \ldots, x(t_n)),$$

i.e., $P(t_1, \ldots, t_n) = (\mathrm{ev}_{t_1}, \ldots, \mathrm{ev}_{t_n})$ and thus $P(t_1, \ldots, t_n)$ is continuous and hence Borel measurable.

- Given $t_1, \ldots, t_n \in [0, 1]$ and $(\alpha_1, \beta_1] \times \cdots \times (\alpha_n, \beta_n] = \prod_{i=1}^{n} (\alpha_i, \beta_i] \subseteq \mathbb{R}^n$, we can define

$$I\left(t_1, \ldots, t_n, \prod_{i=1}^{n} (\alpha_i, \beta_i]\right) = P(t_1, \ldots, t_n)^{-1} \left(\prod_{i=1}^{n} (\alpha_i, \beta_i]\right)$$

$$= \left\{ x \in C_0([0, 1]) \, \middle| \, (x(t_1), \ldots, x(t_n)) \in \prod_{i=1}^{n} (\alpha_i, \beta_i] \right\}.$$

Moreover, we can easily observe that $I\left(t_1, \ldots, t_n, \prod_{i=1}^{n}(\alpha_i, \beta_i]\right)$ is Borel measurable. Furthermore, we note that

$$I\left(t_1, \ldots, t_n, \prod_{i=1}^{n}(\alpha_i, \beta_i]\right) = \bigcap_{i=1}^{n} \mathrm{ev}_{t_i}^{-1}\left((\alpha_i, \beta_i]\right). \qquad (4.2.5)$$

From (4.2.5) it is clear that we can always assume $t_1 \leq t_2 < \cdots < t_{n-1} \leq t_n$.

Exercise 4.2.0.1 Let $t_1, \ldots, t_n \in [0, 1]$ and $t_1 < t_2 \cdots < t_n$. Moreover, let $t_{k-1} < s < t_k$ for $2 \leq k \leq n$. Check that

$$I\left(t_1, \ldots, t_n, \prod_{i=1}^{n}(\alpha_i, \beta_i]\right) = I\left(t_1, \ldots, t_{k-1}, s, t_k, \ldots, t_n, \prod_{i=1}^{k-1}(\alpha_i.\beta_i] \times \mathbb{R} \times \prod_{j=k}^{n}(\alpha_j, \beta_j]\right).$$

Hint: use that $I = \bigcap_{i=1}^{n} \mathrm{ev}_{t_i}^{-1}((\alpha_i, \beta_i])$.

Let us now denote by $\mathcal{I}$ be the collection of all $I\left(t_1, \ldots, t_n, \prod_{i=1}^{n}(\alpha_i, \beta_i]\right)$, where[1] $n \in \mathbb{N}$ and $\alpha_i \leq \beta_i$ with $\alpha_i, \beta_i \in \mathbb{R} \cup \{\infty\}$ for $1 \leq i \leq n$.

Exercise 4.2.0.2 Check that $\mathcal{I}$ is a *semi-algebra*, i.e., satisfies the following axioms:

(1) $\varnothing, C_0([0, 1]) \in \mathcal{I}$,
(2) If $I, J \in \mathcal{I}$, then $I \cap J \in \mathcal{I}$,
(3) If $I \in \mathcal{I}$, then $C_0([0, 1]) \setminus I$ is a finite disjoint union of elements in $\mathcal{I}$.

Solution to Exercise 4.2.0.2

(1) Note that $\varnothing = \mathrm{ev}_1^{-1}((1, 1])$, and thus $\varnothing \in \mathcal{I}$.
(2) Let $I = \bigcap_{i=1}^{n} \mathrm{ev}_{t_i}^{-1}((\alpha_i, \beta_i])$ and $J = \bigcap_{j=1}^{m} \mathrm{ev}_{s_j}^{-1}((\gamma_j, \delta_j])$. Then

$$I \cap J = \bigcap_{\substack{1 \leq i \leq n \\ 1 \leq j \leq m}} \left(\mathrm{ev}_{t_i}^{-1}((\alpha_i, \beta_i]) \cap \mathrm{ev}_{s_j}^{-1}((\gamma_j, \delta_j])\right).$$

Note that $\mathrm{ev}_{t_i}^{-1}((\alpha_i, \beta_i]) \cap \mathrm{ev}_{s_j}^{-1}((\gamma_j, \delta_j])$ is of the form $\mathrm{ev}_t^{-1}((a, b])$.
(3) Let $I \in \mathcal{I}$ such that $I = \mathrm{ev}_t^{-1}((\alpha, \beta])$. Then we get that

$$C_0([0, 1]) \setminus I = \mathrm{ev}_t^{-1}((-\alpha, \alpha]) \cup \mathrm{ev}_t^{-1}((\beta, \alpha]) \in \mathcal{I}.$$

We leave the general case to the reader.

□

---

[1] Note that for us zero is always included in $\mathbb{N}$.

Theorem 4.2.0.1 (Wiener 1923). *There is a unique probability measure μ on $\mathcal{B}(C_0([0, 1]))$, such that*

$$\mu\left(I\left(t_1, ..., t_n, \prod_{i=1}^{n}(\alpha_i, \beta_i]\right)\right) = \frac{\displaystyle\int_{\prod_{i=1}^{n}(\alpha_i, \beta_i]} e^{-\frac{1}{2}\sum_{i=1}^{n}\frac{(x_i - x_{i-1})^2}{t_i - t_{i-1}}} \, dx_1 \cdots dx_n}{\sqrt{(2\pi)^n t_1 (t_2 - t_1) \cdots (t_n - t_{n-1})}}. \tag{4.2.6}$$

Before we prove this theorem, we want to give a small overview of the proof strategy:

- First, we will define $\mu(I)$ for $I \in \mathcal{I}$ by (4.2.6).
- Then we will use the *Caratheodory extension* construction.

Given $I\left(t_i, , , .., t_n, \prod_{i=1}^{n}(\alpha_i, \beta_i]\right)$, define $\mu(I)$ as in (4.2.6). First, we show that μ is well defined i.e., if $t_{k-1} < s < t_k$ for $2 \leq k \leq n$, we get that

$$\mu\left(I\left(t_1, ..., t_n, \prod_{i=1}^{n}(\alpha_i, \beta_i]\right)\right) = \mu\left(I\left(t_1, ..., t_{k-1}, s, t_k, ..., t_n, \prod_{i=1}^{k-1}(\alpha_i, \beta_i] \times \mathbb{R} \times \prod_{i=k}^{n}(\alpha_i, \beta_i]\right)\right). \tag{4.2.7}$$

To verify (4.2.7), we need the following lemma.

Lemma 4.2.0.1 (Kolmogorov–Chapman equation).
Define $K(t, x, y) := \frac{1}{\sqrt{2\pi t}} e^{-\frac{(x-y)^2}{2t}}$. Then we get that

$$\int_{\mathbb{R}} K(t_1, x, y) K(t_2, y, z) dy = K(t_1 + t_2, x, z). \tag{4.2.8}$$

In other words, we get that

$$\frac{1}{\sqrt{(2\pi)^2 t_1 t_2}} \int_{\mathbb{R}} e^{-\frac{(x-y)^2}{2t_1}} e^{-\frac{(y-z)^2}{2t_2}} dy = \frac{1}{\sqrt{2\pi(t_1 + t_2)}} e^{-\frac{(x-z)^2}{2(t_1+t_2)}}. \tag{4.2.9}$$

Proof of Theorem 4.2.0.1 Note that we have

$$\mu\left(I\left(t_1, ..., t_{k-1}, s, t_k, ..., t_n, \prod_{i=1}^{k-1}(\alpha_i, \beta_i] \times \mathbb{R} \times \prod_{i=k}^{n}(\alpha_i, \beta_i]\right)\right) = \tag{4.2.10}$$

$$= \frac{\displaystyle\int_{\prod_{i=1}^{k-1}(\alpha_i, \beta_i] \times \mathbb{R} \times \prod_{i=k}^{n}(\alpha_i, \beta_i]} e^{-\frac{1}{2}\left\{\sum_{i=1}^{k-2}\frac{(x_i - x_{i-1})^2}{t_i - t_{i-1}} + \sum_{i=k+1}^{n}\frac{(x_i - x_{i-1})^2}{t_i - t_{i-1}} + \frac{(y - x_{k-1})^2}{s - t_{k-1}} + \frac{(x_k - y)^2}{t_k - s}\right\}} \, dx_1 \cdots dx_{k-1} dy dx_k \cdots dx_n}{\sqrt{(2\pi)^{n+1} t_1 (t_2 - t_1) \cdots (t_{k-1} - t_{k-2})(s - t_{k-1})(t_k - s) \cdots (t_n - t_{n-1})}}.$$

Using Lemma 4.2.0.1, we see that

$$\frac{\int_{\prod_{i=1}^{n}(\alpha_i,\beta_i]} e^{-\frac{1}{2}\sum_{i=1}^{n}\frac{(x_i-x_{i-1})^2}{t_i-t_{i-1}}} dx_1\cdots dx_n}{\sqrt{(2\pi)^{n+1}t_1(t_2-t_1)\cdots(t_k-t_{k-1})\cdots(t_n-t_{n-1})}} = \mu\left(I\left(t_1,\ldots,t_n,\prod_{i=1}^{n}(\alpha_i,\beta_i]\right)\right).$$

$$(4.2.11)$$

$\square$

Exercise 4.2.0.3 Check that if $I, J \in \mathcal{I}$ and $I \cap J = \varnothing$, $I \cup J \in \mathcal{I}$, then

$$\mu(I \cup J) = \mu(I) + \mu(J),$$

i.e., μ is finitely additive. Hint: Use

$$I = I\left(t_1,\ldots,t_n,\prod_{i=1}^{n}(\alpha_i,\beta_i]\right),$$

$$J = J\left(s_1,\ldots,s_n,\prod_{j=1}^{m}(\gamma_j,\delta_j]\right).$$

A fact of the construction is that μ is countably additive on $\mathcal{I}$. Now, by the Caratheodory extension construction, μ induces a unique measure on $\sigma(\mathcal{I})$, the sigma algebra generated by $\mathcal{I}$. We will denote this "measure" again by μ. To prove Theorem 4.2.0.1, we will show that $\sigma(\mathcal{I}) = \mathcal{B}(C_0([0,1]))$, which is the content of the following proposition.

Proposition 4.2.0.1 *Let everything be defined as before. Then we get that*

$$\sigma(\mathcal{I}) = \mathcal{B}(C_0([0,1])).$$

Proof We already know that $\mathcal{I} \subset \mathcal{B}(C_0([0,1]))$. Hence, $\sigma(\mathcal{I}) \subset \mathcal{B}(C_0([0,1]))$. To show the converse, it suffices to show that for any $\delta > 0$, we have

$$\overline{B_\delta(x_0)} := \{x \in C_0([0,1]) \mid \|x - x_0\| \le \delta\} \subset \sigma(\mathcal{I}).$$

Consider some $\delta > 0$ and $x_0 \in C_0([0,1])$. Our goal will be to show that

$$\overline{B_\delta(x_0)} = \bigcap_{N=1}^{\infty} K_N,$$

where $K_N \in \sigma(\mathcal{I})$ for all N. Note that for some fixed $t \in [0, 1]$ we have

$$\overline{B_\delta(x_0)} \subset \{x \in C_0([0, 1]) \mid \|x(t) - x_0(t)\| \leq \delta\}. \tag{4.2.12}$$

Let $\{t_k\}_{k=1}^{\infty}$ be a dense subset of $[0, 1]$ and define

$$K_N = \{x \in C_0([0, 1]) \mid \|x(t_j) - x_0(t_j)\| \leq \delta \text{ for } j = 1, 2, ..., N\}.$$

Then by (4.2.12), $\overline{B_\delta(x_0)} \subset \bigcap_{N=1}^{\infty} K_N$. To show the reverse inclusion, we will show that

$$x \notin \overline{B_\delta(x_0)} \Longrightarrow x \notin \bigcap_{N=1}^{\infty} K_N.$$

Assume that $x \notin \overline{B_\delta(x_0)}$. Then there is an $s \in [0, 1]$ such that

$$\|x(s) - x_0(s)\| \geq \delta + \delta_1$$

for some $\delta_1 > 0$. Now, choose a subsequence $\{t_{k_j}\}$ of $\{t_k\}$ such that $t_{k_j} \to s$, which can be done since $\{t_k\}$ is dense. Since now x and x_0 are both continuous, we get

$$x(t_{k_j}) \longrightarrow x(s), \tag{4.2.13}$$
$$x_0(t_{k_j}) \longrightarrow x_0(s). \tag{4.2.14}$$

Thus, for large j we get

$$\|x(t_{k_j}) - x_0(t_{k_j})\| \geq \delta + \frac{\delta_1}{2},$$

and thus $x \notin \bigcap_{N=1}^{\infty} K_N$. Hence, we were able to construct a measure on $\mathcal{B}(C_0([0, 1]))$. To complete the proof of Theorem 4.2.0.1, we check that μ is a probability measure. Indeed, we have that

$$\mu(C_0([0, 1])) = \mu(\mathrm{ev}_1^{-1}(\mathbb{R})) = \frac{1}{\sqrt{2\pi}} \int_{\mathbb{R}} e^{-\frac{x^2}{2}} dx = 1.$$

This completes the proof of Theorem 4.2.0.1.   □

Let us now compute the Wiener measure of the set

$$A_{s,t}^{a,b} := \{x \in C_0([0, 1]) \mid a \leq x(t) - x(s) \leq b\},$$

where $a, b \in \mathbb{R}$ with $a \leq b$, and $s, t \in [0, 1]$ with $0 \leq s < t \leq 1$. Note that $A_{s,t}^{a,b} = P(s, t)^{-1}(E)$, where $E = \{(x, y) \in \mathbb{R}^2 \mid a \leq x - y \leq b\}$. Hence, we get that

$$\mu(A_{s,t}^{a,b}) = \frac{1}{\sqrt{(2\pi)^2 s(t-s)}} \iint_E e^{-\frac{y^2}{2s} - \frac{(x-y)^2}{2(t-s)}} dxdy$$

$$= \frac{1}{\sqrt{(2\pi)^2 s(t-s)}} \int_{\mathbb{R}} \left(\int_a^b e^{-\frac{u^2}{2(t-s)}} du \right) e^{-\frac{y^2}{2s}} dy \qquad (4.2.15)$$

$$= \frac{1}{\sqrt{2\pi(t-s)}} \int_a^b e^{-\frac{u^2}{2(t-s)}} du.$$

We want to make a short input on the *pushforward* of a measure. Let $(X, \sigma(X), \mu)$ be a measure space, $(Y, \sigma(Y))$ a measurable space and $f : X \to Y$ a measurable map. Then we can define a measure $f_*\mu$ on $(Y, \sigma(Y))$, which is defined as

$$f_*\mu(P) = \mu(f^{-1}(P)), \qquad P \in \sigma(Y).$$

This measure $f_*\mu$ is called the pushfoward measure of μ along f. It is easy to check that for any integrable function $\alpha \colon Y \to \mathbb{R}$ we have

$$\int_Y \alpha(y) d(f_*\mu(y)) = \int_X (f^*\alpha)(x) d\mu(x),$$

where we define $(f^*\alpha)(x) := \alpha(f(x))$. Define the map $\alpha_{s,t} \colon C_0([0, 1]) \to \mathbb{R}$ by $\alpha_{s,t}(x) = x(t) - x(s)$. Then (4.2.15) implies that $(\alpha_{s,t})_*\mu$, where μ is the Wiener measure on $C_0([0, 1])$, is given by

$$(\alpha_{s,t})_*\mu([a, b]) = \frac{1}{\sqrt{2\pi(t-s)}} \int_a^b e^{-\frac{x^2}{2(t-s)}} dx.$$

Thus $(\alpha_{s,t})_*\mu$ is the centered *Gaussian measure* on $\mathbb{R}$ with variance $(t - s)$. This discussion leads to the following corollary.

Corollary 4.2.0.1 *Let everything be as before. Then the following hold:*

$$\int_{C_0([0,1])} (x(t) - x(s)) d\mu(x) = 0, \qquad (4.2.16)$$

$$\int_{C_0([0,1])} (x(t) - x(s))^2 d\mu(x) = t - s. \qquad (4.2.17)$$

Exercise 4.2.0.4 Show that

$$\int_{C_0([0,1])} x(s) x(t) d\mu(x) = \min_{s,t \in [0,1]} \{s, t\}.$$

Hint: Assume that $s < t$ and show that

$$\frac{1}{\sqrt{2\pi s(t-s)}} \iint_{\mathbb{R}^2} xy e^{-\frac{x^2}{2s}} e^{-\frac{(x-y)^2}{2(t-s)}} dxdy = s.$$

Exercise 4.2.0.5 Compute the integrals

$$\int_{C_0([0,1])} \left(\int_0^1 x(t)dt \right) d\mu(x), \tag{4.2.18}$$

$$\int_{C_0([0,1])} \left(\int_0^1 x(t)^2 dt \right) d\mu(x). \tag{4.2.19}$$

Hint: Use Fubini's theorem.

4.2.1   Towards Nowhere Differentiability of Brownian Paths

Let h be some positive number, $\alpha \in (0, 1]$, and define the following spaces:

$$C_h^\alpha(s, t) := \{x \in C_0([0, 1]) \mid \|x(t) - x(s)\| \leq h|t - s|^\alpha\},$$
$$C_h^\alpha(t) := \bigcap_{s \in [0,1]} C_h^\alpha(s, t),$$
$$C_h^\alpha := \bigcap_{t \in [0,1]} C_h^\alpha(t).$$

One can actually check that $C_h^\alpha(s, t)$ is closed in $C_0([0, 1])$ and thus the sets $C_h^\alpha(s, t)$, $C_h^\alpha(t)$ and C_h^α are Borel measurable.

Lemma 4.2.1.1 *Let everything be as before and denote by μ the Wiener measure. Then we get that*

$$\mu(C_h^\alpha(s, t)) \leq \sqrt{\frac{2}{\pi}} h|t - s|. \tag{4.2.20}$$

Proof Note that we can write

$$C_h^\alpha(s, t) = \{x \in C_0([0, 1]) \mid -h|t - s|^\alpha \leq x(t) - x(s) \leq h|t - s|^\alpha\} =: A_{s,t}^{-h|t-s|^\alpha, h|t-s|^\alpha}.$$

Assume now that $s < t$. Then, by (4.2.15), we have

$$\mu(C_h^\alpha(s, t)) = \frac{1}{\sqrt{2\pi}} \int_{-h|t-s|^\alpha}^{h|t-s|^\alpha} e^{-\frac{u^2}{2(t-s)}} du$$

$$= \frac{1}{\sqrt{2\pi}} \int_{-h|t-s|^{\alpha-\frac{1}{2}}}^{h|t-s|^{\alpha-\frac{1}{2}}} e^{-\frac{u^2}{2}} du \tag{4.2.21}$$

$$\leq \sqrt{\frac{2}{\pi}} h|t - s|^{\alpha-\frac{1}{2}},$$

where we used that $e^{-\frac{u^2}{2}} \leq 1$. $\qquad\square$

Corollary 4.2.1.1 *If $\frac{1}{2} < \alpha \leq 1$, then $\mu(C_h^\alpha(t)) = 0$ and hence $\mu(C_h^\alpha) = 0$.*

Proof Let $\{t_k\} \subseteq [0, 1]$ such that $t_k \xrightarrow{k\to\infty} t$. Then it is not hard to see that $C_h^\alpha(t) \subseteq C_h^\alpha(t, t_k)$. Thus, we get that

$$\mu(C_h^\alpha(t)) \leq \mu(C_h^\alpha(t, t_k)) \leq \sqrt{\frac{2}{\pi}} h |t - t_k|^{\alpha - \frac{1}{2}} \xrightarrow{k\to\infty} 0.$$

$\square$

Proposition 4.2.1.1 *If $\frac{1}{2} < \alpha \leq 1$, we get that*

$$\mu(\{x \in C_0([0, 1]) \mid x \text{ is Hölder continuous with exponent } \alpha\}) = 0.$$

Proof The proof is straightforward since we clearly have that

$$\{x \in C_0([0, 1]) \mid x \text{ is Hölder continuous with exponent } \alpha\} \subseteq \bigcup_{h=1}^{\infty} C_h^\alpha.$$

$\square$

Corollary 4.2.1.2 *We have that*

$$\mu(\{x \in C_0([0, 1]) \mid x \text{ is differentiable}\}) = 0.$$

Proof The proof is straightforward since we clearly have that

$$\{x \in C_0([0, 1]) \mid x \text{ is differentiable}\}$$
$$\subseteq \{x \in C_0([0, 1]) \mid x \text{ is Hölder continuous of exponent } 1\}.$$

$\square$

The following lemma will play an important role when we discuss nowhere differentiability of Brownian paths.

Lemma 4.2.1.2 *For $t \in [0, 1]$ we get that $\mu(D_t) = 0$, where $D_t := \{x \in C_0([0, 1]) \mid \dot{x}(t) \text{ exists}\}$.*

Proof We can easily check that $D_t \subseteq \bigcup_{h=1}^{\infty} C_h^\bullet(t)$. Moreover, we already know that $\mu(C_h^\bullet(t)) = 0$ and thus $\mu(D_t) = 0$. $\square$

Lemma 4.2.1.3 *Define a map F on $C_0([0, 1]) \times [0, 1]$ by*

$$F(x, t) := \begin{cases} 1, & \text{if } \dot{x}(t) \text{ exists,} \\ 0, & \text{otherwise.} \end{cases}$$

Then F is measurable on $C_0([0, 1]) \times [0, 1]$.

Proof We will show that the set $G := \{(x, t) \mid F(x, t) = 1\}$ has measure 0 with respect to $\mu \times \lambda$, where λ is the Lebesgue measure on $[0, 1]$. First, we observe that $G \subseteq G^*$, where

$$G^* = \left\{ (x, t) \in C_0([0, 1]) \times [0, 1] \;\middle|\; \lim_{n \to \infty} f_n(x, t) \text{ exists} \right\},$$

with $f_n(x, t) := \frac{x\left(t - \frac{1}{n}\right) - x(t)}{\frac{1}{n}}$. Moreover, G^* is measurable since it is the set where a sequence of measurable functions have a limit. Note that we have

$$(\mu^* \times \lambda)(G) \leq (\mu^* \times \lambda)(G^*) = (\mu \times \lambda)(G^*),$$

where μ^* is the *outer measure* associated with the premeasure μ in the construction of the Wiener measure. Hence, we get that

$$(\mu \times \lambda)(G^*) = \int_0^1 \mu(G_t^*) \mathrm{d}\lambda(t),$$

where $G_t^* := \left\{ x \in C_0([0, 1]) \;\middle|\; \lim_{n \to \infty} f_n(x, t) \text{ exists} \right\}$. If we can show that $\mu(G_t^*) = 0$, then we see that $(\mu \times \lambda)(G^*) = 0$ which means that $(\mu^* \times \lambda)(G) = 0$ and thus G is measurable. To see that $\mu(G_t^*) = 0$, one can show that

$$G_t^* \subseteq \bigcup_{h=1}^{\infty} \bigcap_{n=1}^{\infty} C_h^\bullet \left(t, t + \frac{1}{n} \right),$$

and use the fact that $\lim_{n \to \infty} \mu\left(C_h^\bullet \left(t, t + \frac{1}{n} \right) \right) = 0$. $\square$

Theorem 4.2.1.1 (Nowhere differentiable Brownian paths). *With sure probability, paths $x \in C_0([0, 1])$ are at most differentiable on a subset of Lebesgue measure 0 on $[0, 1]$. In other words: with probability 1, paths $x \in C_0([0, 1]))$ are "nowhere" differentiable.*

Proof Let F be the map defined as in Lemma 4.2.1.3. Note that, by Lemma 4.2.1.2, we then have

$$\int_{C_0([0,1]) \times [0,1]} F(x, t) \mathrm{d}\mu(x) \mathrm{d}t = \int_0^1 \left(\int_{C_0([0,1])} F(x, t) \mathrm{d}\mu(x) \right) \mathrm{d}t = \int_0^1 \mu(D_t) \mathrm{d}t = 0,$$

$$(4.2.22)$$

since for fixed t, we have[2] $F(x, t) = \chi_{D_t}$. Thus, we get that

---

[2] Here χ_{D_t} denotes the characteristic function of the set D_t.

$$\int_{C_0([0,1])} \left(\int_0^1 F(x,t) dt \right) d\mu(x) = 0,$$

whence $\int_0^1 F(x,t) dt = 0$ for almost all $x \in C_0([0,1])$. For such x we get that $F(x,t) = 0$ for almost all $t \in [0,1]$ and thus $\dot{x}(t)$ does not exist for almost all $t \in [0,1]$. □

The following facts are stated without proof.

- We have that $\mu(\{x \in C_0([0,1]) \mid x \text{ is Hölder continuous with exponent } \alpha\}) = 1$ for $0 \le \alpha \le \frac{1}{2}$ (cf. Kuo (1975)).
- We have that $\mu(\{x \in C_0([0,1]) \mid x \text{ is Hölder continuous with exponent } \frac{1}{2}\}) = 0$ (cf. Simon (1979); Mörters and Peres (2010)).

Remark 4.2.1.1 More generally, we can also consider the Wiener measure μ_x on the set

$$C_x([a,b]) := \{\omega \colon [a,b] \to \mathbb{R} \mid \omega \text{ is continuous and } \omega(a) = x\}.$$

Moreover, can consider the Wiener measure μ_x^y on

$$C_x^y([a,b]) := \{\omega \colon [a,b] \to \mathbb{R} \mid \omega \text{ is continuous and } \omega(a) = x, \omega(b) = y\}.$$

This measure is the unique measure on the Borel sigma algebra $\mathcal{B}(C_x^y([0,1]))$ such that for all $t_1, ..., t_n \in (a,b)$ we get that

$$\mu_x^y(I(t_1, ..., t_n), E) = \int_E K_{b-t_n}(y, x_n) K_{t_n - t_{n-1}}(x_n, x_{n-1}) \cdots K_{t_1 - a}(x_1, x) dx_1 \cdots dx_n,$$

where $E \subseteq \mathbb{R}^n$ is a measurable set and $K_t(u,v) = \frac{1}{\sqrt{2\pi t}} e^{-\frac{1}{2} \frac{(u-v)^2}{t}}$.

Definition 4.2.1.1 (*Conditional Wiener measure*). The Wiener measure μ_x^y is called a *conditional Wiener measure*.[3]

Remark 4.2.1.2 The conditional Wiener measure μ_x^y is *not* a probability measure. In fact, we have that

$$\mu_x^y(C_x^y([a,b])) = \frac{1}{\sqrt{2\pi(b-a)}} e^{-\frac{(x-y)^2}{t(b-a)}}.$$

---

[3] The measure μ_x^y is actually called the *conditional Wiener measure* because μ_x and μ_x^y fits in the general framework of a conditional measure (cf. Kuo (1975); Mörters and Peres (2010)). Thus, we have

$$\mu_x = \int_{\mathbb{R}} \mu_x^y dy.$$

4.2.2 The Feynman–Kac Formula

The goal of this subsection is to prove the following theorem, which relates a *Wick rotated*, i.e., a Euclidean, version of Feynman's path integral to the concept of conditional expectation.

Theorem 4.2.2.1 (Feynman–Kac (Kac 1949; Glimm and Jaffe 1987)). *Let V be a continuous function on $\mathbb{R}$, which is bounded from below. Consider the Hamiltonian operator $\hat{H} = \hat{H}_0 + V(\hat{x})$ with kinetic term given by $\hat{H}_0 = -\frac{1}{2}\frac{d^2}{dx^2} = -\frac{1}{2}\Delta$. Moreover, assume that $\hat{H}$ is essentially self-adjoint. Then for all $\psi \in L^2(\mathbb{R})$ we get that*

$$\left(e^{-t\hat{H}}\psi\right)(x_0) = \int_{C_{x_0}([0,t])} \psi(x(t)) e^{-\int_0^t V(x(s))ds} d\mu_{x_0}(x). \tag{4.2.23}$$

The main technical tool, which we are going to use here, is the KLT product formula, as we have seen before, given in the following way. Let A and B be self-adjoint operators bounded from below on some Hilbert space $\mathcal{H}$. Assume that $\hat{H} = A + B$ is essentially self-adjoint on the intersection of the domains of A and B, i.e., on $D(A) \cap D(B)$. We will denote the unique self-adjoint extension of $\hat{H}$ again by $\hat{H}$. Then for all $\phi \in \mathcal{H}$ and for all $t \geq 0$ we get that

$$e^{-t\hat{H}}\phi = \lim_{n\to\infty} \left(\left(e^{-\frac{t}{n}A}e^{-\frac{t}{n}B}\right)^n \phi\right).$$

Proof of Theorem 4.2.2.1 Recall first that

$$\left(\left(e^{-\frac{t}{n}\hat{H}_0}e^{-\frac{t}{n}V}\right)\psi\right)(x_0) = \int_{\mathbb{R}} K_{\frac{t}{n}}(x_1, x_0)e^{-\frac{t}{n}V(x_1)}\psi(x_1)dx_1. \tag{4.2.24}$$

Taking the square of this operator we get

$$\left(\left(e^{-\frac{t}{n}\hat{H}_0}e^{-\frac{t}{n}V}\right)^2\psi\right)(x_0) = \iint_{\mathbb{R}^2} K_{\frac{t}{n}}(x_2, x_1)K_{\frac{t}{n}}(x_1, x_0)e^{-\frac{t}{n}(V(x_2)+V(x_1))}\psi(x_1)dx_1 dx_2. \tag{4.2.25}$$

Thus, taking the n-th power of the operator, we get

$$\left(\left(e^{-\frac{t}{n}\hat{H}_0}e^{-\frac{t}{n}V}\right)^n\psi\right)(x_0) = \int_{\mathbb{R}^n} K_{\frac{t}{n}}(x_n, x_{n-1})\cdots K_{\frac{t}{n}}(x_1, x_0)e^{-\frac{t}{n}\sum_{j=1}^n V(x_j)}\psi(x_n)dx_1\cdots dx_n, \tag{4.2.26}$$

where $x_j = x\left(\frac{jt}{n}\right)$ and thus $x_n = x(t)$. It is easy to see that (4.2.26) is actually equal to

$$\int_{C_{x_0}([0,t])} \psi(x(t))e^{-\frac{t}{n}\sum_{j=1}^n V\left(x\left(\frac{jt}{n}\right)\right)} d\mu_{x_0},$$

and thus we get that

$$\left(e^{-t\hat{H}}\psi\right)(x_0) = \lim_{n\to\infty} \int_{C_{x_0}([0,t])} \psi(x(t)) e^{-\frac{t}{n}\sum_{j=1}^{n} V\left(x\left(\frac{jt}{n}\right)\right)} d\mu_{x_0}. \tag{4.2.27}$$

Since the limit for $n \to \infty$ of the exponential is given by

$$\lim_{n\to\infty} e^{-\frac{t}{n}\sum_{j=1}^{n} V\left(x\left(\frac{jt}{n}\right)\right)} = e^{-\int_0^1 V(x(s))ds},$$

it is enough to justify that we can pull the limit inside the integral in (4.2.27). This can be easily justified by using the assumption that V is bounded from below and by Lebesgue's dominated convergence theorem. We leave the details to the reader. □

Remark 4.2.2.1 In fact, the Feynman–Kac formula holds for some general V (cf. Simon (1979)).

Remark 4.2.2.2 There is also a Feynman–Kac formula with respect to the conditional Wiener measure μ_x^y on $C_x^y([0, t])$ (cf. Glimm and Jaffe (1987)). It simply says that the integral kernel of $e^{-t\hat{H}}$ is given by

$$K_t(x, y, \hat{H}) = \int_{C_x^y([0,t])} e^{-\int_0^1 V(x(s))ds} d\mu_x^y.$$

4.3   Gaussian Measures

4.3.1   Gaussian Measures on $\mathbb{R}$

Definition 4.3.1.1 (*Gaussian measure I*). A Borel probability measure μ on $\mathbb{R}$ is called *Gaussian* if it is either the Dirac measure δ_a at some $a \in \mathbb{R}$, or it is of the form

$$d\mu(x) = \frac{1}{\sqrt{2\pi\sigma}} e^{-\frac{(x-a)^2}{2\sigma}} dx, \tag{4.3.1}$$

for some $a \in \mathbb{R}$, and $\sigma > 0$. The parameters a and σ are usually called *mean* and *variance* of the measure μ, respectively.

Remark 4.3.1.1 If μ is given by (4.2.4), we say μ is *non-degenerate Gaussian measure*. Moreover, if $a = 0$, then μ is called a *centered Gaussian measure*.

Exercise 4.3.1.1 Check that

$$a = \int_{\mathbb{R}} x d\mu(x), \tag{4.3.2}$$

$$\sigma = \int_{\mathbb{R}} (x - a)^2 \mathrm{d}\mu(x).\tag{4.3.3}$$

Exercise 4.3.1.1 justifies the names "mean" and "variance" of the Gaussian measure μ given by (4.2.4).

Exercise 4.3.1.2 Given a Borel measure μ, we can define a map $\hat{\mu}\colon \mathbb{R} \to \mathbb{C}$ by

$$\hat{\mu}(y) := \int_{\mathbb{R}} \mathrm{e}^{\mathrm{i}yx} \mathrm{d}\mu(x).$$

Check that if μ is given by (4.2.4), we get $\hat{\mu}(y) = \mathrm{e}^{\mathrm{i}ay - \frac{1}{2}\sigma y^2}$.

Definition 4.3.1.2 (*Characteristic Functional I*). The map $\hat{\mu}$ defined as in Exercise 4.3.1.2 is called the *characteristic functional* or *Fourier transform* of the measure μ.

Exercise 4.3.1.3 Let μ be a Borel measure on $\mathbb{R}$. Show that μ is Gaussian if and only if

$$\hat{\mu}(y) = \mathrm{e}^{\mathrm{i}ay - \frac{1}{2}\sigma y^2}\tag{4.3.4}$$

for some $a \in \mathbb{R}$ and $\sigma > 0$.

4.3.2 Gaussian Measures on Finite-Dimensional Vector Spaces

Definition 4.3.2.1 (*Gaussian measure II*). A Borel probability measure μ on $\mathbb{R}^n$ is called *Gaussian*, if for all linear maps $\alpha\colon \mathbb{R}^n \to \mathbb{R}$, the pushforward measure $\alpha_*\mu$ is Gaussian on $\mathbb{R}$.

This definition is quite abstract, but we will later give a more "working" definition of a Gaussian measure.

Remark 4.3.2.1 From now on we will identify the dual space $(\mathbb{R}^n)^*$ with $\mathbb{R}^n$, using the standard metric on $\mathbb{R}^n$, i.e., a linear map $\alpha\colon \mathbb{R}^n \to \mathbb{R}$ will be considered as a vector $\alpha \in \mathbb{R}^n$.

Definition 4.3.2.2 (*Characteristic functional II*). Given a finite Borel measure μ on $\mathbb{R}^n$, we can define a map $\hat{\mu}\colon \mathbb{R} \to \mathbb{C}$ by

$$\hat{\mu}(y) := \int_{\mathbb{R}} \mathrm{e}^{\mathrm{i}\langle y, x \rangle} \mathrm{d}\mu(x).$$

The map $\hat{\mu}$ is called the *characteristic functional* or *Fourier transform* of the measure μ.

Proposition 4.3.2.1 *A Borel measure μ on $\mathbb{R}^n$ is Gaussian if and only if*

$$\hat{\mu}(y) = e^{-i\langle y,a\rangle - \frac{1}{2}\langle Ky,y\rangle}, \tag{4.3.5}$$

where $a \in \mathbb{R}^n$ and K is a positive-definite symmetric $n \times n$-matrix. In this case, when μ is non-degenerate, then

$$d\mu(x) = \frac{1}{\det\left(\frac{K}{2\pi}\right)^{\frac{1}{2}}} e^{-\frac{1}{2}\langle K^{-1}(x-a), K^{-1}(x-a)\rangle} dx.$$

Proof Given a Borel measure μ on $\mathbb{R}^n$ and a linear map $\alpha \colon \mathbb{R}^n \to \mathbb{R}$, we get that

$$\widehat{\alpha_*\mu}(t) = \int_{\mathbb{R}} e^{its} d(\alpha_*\mu)(s) = \int_{\mathbb{R}} e^{it\alpha(x)} d\mu(x)$$
$$= \int_{\mathbb{R}^n} e^{i\langle t\alpha, x\rangle} d\mu(x) \tag{4.3.6}$$
$$= \hat{\mu}(t\alpha),$$

where in the second equality we have used that

$$\int_X (f^*\alpha)(x) dx = \int_Y f(y) d(\alpha_*\mu)(y).$$

Assume that $\hat{\mu}$ has the form (4.3.4). Then we get that

$$\widehat{\alpha_*\mu}(t) = \hat{\mu}(t\alpha) = e^{i\langle t\alpha, a\rangle - \frac{1}{2}\langle K(t\alpha), t\alpha\rangle} = e^{it\langle \alpha, a\rangle - \frac{1}{2}t^2\langle K\alpha, \alpha\rangle}. \tag{4.3.7}$$

By Exercise 4.3.1.2, we get that $\widehat{\alpha_*\mu}$ is a Gaussian measure on $\mathbb{R}$. Conversely, assume that $\alpha_*\mu$ is a Gaussian measure on $\mathbb{R}$ for all linear maps $\alpha \colon \mathbb{R}^n \to \mathbb{R}$. By Exercise 4.3.1.2, we get

$$\widehat{\alpha_*\mu}(t) = e^{ita(\alpha) - \frac{1}{2}\sigma(\alpha)t^2}.$$

Moreover, by Exercise 4.3.1.1, we get

$$a(\alpha) = \int_{\mathbb{R}} t \, d(\alpha_*\mu)(t), \tag{4.3.8}$$

$$\sigma(\alpha) = \int_{\mathbb{R}} (t - a(\alpha))^2 d(\alpha_*\mu)(t). \tag{4.3.9}$$

We can check that the map $\alpha \mapsto a(\alpha)$ defines a linear map $\mathbb{R}^n \to \mathbb{R}$, and hence it can be identified with $a \in \mathbb{R}^n$ as $a(\alpha) = \langle a, \alpha\rangle$. Moreover, the map $\alpha \mapsto \sigma(\alpha)$ defines a quadratic form on $\mathbb{R}^n$. Hence, there is a symmetric $n \times n$-matrix K such that $\sigma(\alpha) = \langle K\alpha, \alpha \rangle$. Thus, $\sigma(\alpha) > 0$ for all $\alpha \in \mathbb{R}^n$ implies that K is a positive matrix. The last part of the proof is left as an exercise.[4]   □

In conclusion, we have seen that this abstract definition of a Gaussian measure on $\mathbb{R}^n$ is equivalent to the usual notion of Gaussian measure.

Exercise 4.3.2.1 Let μ be a Gaussian measure on $\mathbb{R}^n$ of the form

$$d\mu(x) = \det \left(\frac{K}{2\pi} \right)^{\frac{1}{2}} e^{-\frac{1}{2}\langle K(x-a),(x-a)\rangle} dx.$$

Check that

$$a = \int_{\mathbb{R}} x d\mu(x) = \left(\int_{\mathbb{R}} x_1 d\mu(x_1), ..., \int_{\mathbb{R}} x_n d\mu(x_n) \right),$$

and

$$K_{ij}^{-1} = \int_{\mathbb{R}^n} (x_i - a_i)(x_j - a_j) d\mu(x).$$

Definition 4.3.2.3 (*Covariance operator*). The vector $a \in \mathbb{R}^n$ is called the *mean* of the Gaussian measure and the matrix K^{-1} is called the *covariance operator* of μ. When the mean of a Gaussian measure is 0, then it is called a *centered Gaussian* (see Remark 4.3.1.1).

Proposition 4.3.2.2 *Let μ be a centered Gaussian measure on $\mathbb{R}^n$ of the form*

$$d\mu(x) = \det \left(\frac{K}{2\pi} \right)^{\frac{1}{2}} e^{-\frac{1}{2}\langle Kx,x\rangle} dx.$$

Then the following hold:

(1)   For all $\lambda \in \mathbb{C}^n$, we have that

$$\int_{\mathbb{R}^n} e^{\langle \lambda,x\rangle} d\mu(x) = e^{\frac{1}{2}\langle K^{-1}\lambda,\lambda\rangle}.$$

(2)   We have that

$$\int_{\mathbb{R}^n} f(x - \sqrt{t}y) d\mu(y) = \left(e^{\frac{t}{n} L^\mu} f \right)(x),$$

where $L^\mu := \sum_{i=1}^{n} (K^{-1})_{ij} \frac{\partial}{\partial x_i} \frac{\partial}{\partial y_j}$. Moreover, we have that

$$\int_{\mathbb{R}^n} f(y) d\mu(y) = \left(e^{\frac{1}{2} L^\mu} f \right)(0).$$

---

[4] It essentially follows from the 1-dimensional case and diagonalization of K.

(3) We have that

$$\int_{\mathbb{R}^n} p(x)d\mu(x) = p(D_\lambda)e^{-\frac{1}{2}\langle A^{-1}\lambda,\lambda\rangle}\Big|_{\lambda=0},$$

where $p(D_\lambda)$ is a polynomial in derivatives $\frac{\partial}{\partial\lambda_i}$ in λ_i-directions corresponding to the polynomial map $p(x)$, i.e., if e.g., $p(x) = x_1 x_2$, then $p(D_\lambda) = \frac{\partial}{\partial\lambda_1}\frac{\partial}{\partial\lambda_2}$.

Proof

(1) Note that we have

$$\int_{\mathbb{R}^n} e^{\langle\lambda,x\rangle}d\mu(x) = \det\left(\frac{K}{2\pi}\right)^{\frac{1}{2}} \int_{\mathbb{R}^n} e^{\langle\lambda,x\rangle}e^{-\frac{1}{2}\langle Kx,x\rangle}dx = \frac{\det\left(\frac{K}{2\pi}\right)^{\frac{1}{2}}}{\det\left(\frac{K}{2\pi}\right)^{\frac{1}{2}}}e^{\frac{1}{2}\langle K^{-1}\lambda,\lambda\rangle} = e^{\frac{1}{2}\langle K^{-1}\lambda,\lambda\rangle}.$$

(2) It is sufficient to check that $f(x)$ is of the form $e^{\langle\lambda,x\rangle}$ with $\lambda \in \mathbb{C}^n$ as these functions form a dense subset. For $f(x) = e^{\langle\lambda,x\rangle}$, we get

$$f(x - \sqrt{t}y) = e^{\langle\lambda,x\rangle}e^{\langle-\sqrt{t}\lambda,y\rangle},$$

and thus

$$\int_{\mathbb{R}^n} f(x - \sqrt{t}y)d\mu(y) = e^{\langle\lambda,x\rangle}\int_{\mathbb{R}^n} e^{\langle-\sqrt{t}\lambda,y\rangle}d\mu(y) = e^{\langle\lambda,x\rangle}e^{\frac{t}{2}\langle K^{-1}\lambda,\lambda\rangle}.$$

On the other hand

$$L^\mu\left(e^{\langle\lambda,x\rangle}\right) = \langle K^{-1}\lambda, \lambda\rangle e^{\langle\lambda,x\rangle},$$

which means that

$$e^{\frac{t}{2}L^\mu}\left(e^{\langle\lambda,x\rangle}\right) = e^{\frac{t}{2}\langle K^{-1}\lambda,\lambda\rangle}e^{\langle\lambda,x\rangle}.$$

Thus, if $f(x) = e^{\langle\lambda,x\rangle}$, we have

$$\int_{\mathbb{R}} f(x - \sqrt{t}y)d\mu(y) = \left(e^{\frac{t}{2}L^\mu}f\right)(x).$$

The second part can be verified in a similar way. We leave this as an exercise for the reader.

(3) We leave the entire proof of this point as an exercise for the reader. □

Example 4.3.2.1 Let us consider integrals of the form

$$\int_{\mathbb{R}} x_i x_j d\mu(x) = \left(K_{ij}^{-1}\right).$$

More generally, we can consider integrals of the form

$$\int_{\mathbb{R}^n} \langle u, x \rangle \langle v, x \rangle \mathrm{d}\mu(x) = \langle K^{-1}u, v \rangle.$$

Such an integral can be interpreted as a graph with two vertices, representing the points u and v, and an edge, representing the operator K^{-1}, connecting them.

$$u \bullet \overset{K^{-1}}{\rule{3cm}{0.4pt}} \bullet v$$

Example 4.3.2.2 Let us consider the even more general integral

$$\int_{\mathbb{R}^n} \left(\prod_{i=1}^{4} \langle u_i, x \rangle \right) \mathrm{d}\mu(x) = \langle K^{-1}u_1, u_2 \rangle \langle K^{-1}u_3, u_4 \rangle + \langle K^{-1}u_1, u_3 \rangle \langle K^{-1}u_2, u_4 \rangle$$

$$+ \langle K^{-1}u_1, u_4 \rangle \langle K^{-1}u_2, u_3 \rangle.$$

It can be represented as the sum of the following three graphs:

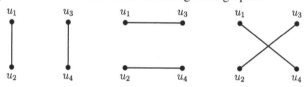

Here each edge represents the operator K^{-1}.

In general, we have the following theorem.

Theorem 4.3.2.1 (Wick 1950). *Let μ be a standard Gaussian measure on $\mathbb{R}^n$. Then we have*

$$\int_{\mathbb{R}^n} \left(\prod_{i=1}^{k} \langle u_i, x \rangle \right) \mathrm{d}\mu = \begin{cases} \sum_{\sigma \in \Pi(s)} \langle K^{-1}u_{\sigma(1)}, u_{\sigma(2)} \rangle \cdots \langle K^{-1}u_{\sigma(2s-1)}, u_{\sigma(2s)} \rangle, & \text{if } k = 2s, \\ 0, & \text{otherwise,} \end{cases}$$

where $\Pi(s)$ denotes the subset of permutations S_{2s} such that $\sigma(2i-1) < \sigma(2i)$ for $i = 1, \ldots, s$ and $\sigma(1) < \sigma(3) < \ldots < \sigma(2s-3) < \sigma(2s-1)$.

Exercise 4.3.2.2 Prove Theorem 4.3.2.1.

4.3.3 Gaussian Measures on Real Separable Hilbert Spaces

Let us now consider the more general case of a real separable Hilbert space $(\mathcal{H}, \langle \cdot, \cdot \rangle_{\mathcal{H}})$.

Definition 4.3.3.1 (*Borel measure on $\mathcal{H}$*). A *Borel measure* μ on $\mathcal{H}$ is a measure which is defined on the Borel sigma algebra $\mathcal{B}(\mathcal{H})$ of $\mathcal{H}$.

In the previous subsection, we saw that a Gaussian measure on a finite-dimensional vector space V is determined by some element $a \in V$ and a positive symmetric matrix K^{-1}, called the *covariance* of the Gaussian mean. In this section we will see whether this is also true for the infinite-dimensional case. Let μ be a Borel measure on $\mathcal{H}$. We define an operator S_μ on $\mathcal{H}$ by

$$\langle S_\mu(x), y \rangle_{\mathcal{H}} := \int_{\mathcal{H}} \langle x, z \rangle_{\mathcal{H}} \langle y, z \rangle_{\mathcal{H}} \mathrm{d}\mu(z). \tag{4.3.10}$$

Remark 4.3.3.1 We have to be a bit careful with this definition. It may happen actually that S_μ does not exist.

Let us now recall some background material on operator theory.

(1) (Trace class operators) Let $A \colon \mathcal{H} \to \mathcal{H}$ be a bounded operator. We define the *square root* of A by the unique positive operator $\sqrt{A}$ with the property

$$A = (\sqrt{A})^* \sqrt{A},$$

which exists by the continuous spectral theorem. Note that $\sqrt{A} \geq 0$. Let A be a non-negative operator on $\mathcal{H}$. Then we get that the sum

$$\sum_{n=1}^{\infty} \langle A e_n, e_n \rangle_{\mathcal{H}}$$

is independent of the choice of an orthonormal basis $\{e_n\}$ of $\mathcal{H}$. In this case, one defines the *trace* of A as

$$\mathrm{Tr}(A) := \sum_{n=1}^{\infty} \langle A e_n, e_n \rangle_{\mathcal{H}}.$$

Definition 4.3.3.2 (*Trace class*). The operator A is called *trace class* if $\mathrm{Tr}(|A|) < \infty$, where $|A|$ denotes the positive square root of A^*A, i.e., $|A| := \sqrt{A^*A}$.

If A is a trace class operator, then $\sum_{n=1}^{\infty} \langle A e_n, e_n \rangle_{\mathcal{H}}$ does not depend on the choice of an orthonormal basis $\{e_n\}$ of $\mathcal{H}$. In this case, we define $\mathrm{Tr}(A) := \sum_{n=1}^{\infty} \langle A e_n, e_n \rangle_{\mathcal{H}}$.

(2) (Bilinear forms/quadratic forms) A *bilinear form* B on $\mathcal{H}$ with domain $D(B)$ is a bilinear map

$$B \colon D(B) \times D(B) \longrightarrow \mathbb{R},$$
$$(x, y) \longmapsto B(x, y),$$

where $D(B)$ is a dense subspace of $\mathcal{H}$. Given a bilinear form B on $\mathcal{H}$, we can define a *quadratic form* $q(x) := B(x, x)$. Moreover, a bilinear form B is called *bounded* if there is some $\varepsilon > 0$ such that for all $x, y \in D(B)$ we get that

$$|B(x, y)| \leq \varepsilon \|x\|_{\mathcal{H}} \|y\|_{\mathcal{H}}.$$

We call B *symmetric*, if $B(x, y) = B(y, x)$ for all $x, y \in \mathcal{H}$. Moreover, B is called *positive(-definite)* if $q(x) \geq 0$, where $q(x) = 0$ if and only if $x = 0$, for all $x \in \mathcal{H}$. If B is a bounded, positive and symmetric bilinear form, then there is a bounded linear operator $S_B \colon \mathcal{H} \to \mathcal{H}$ such that $B(x, y) = \langle S_B(x), y \rangle_{\mathcal{H}}$.

Next, we want to investigate in what cases S_μ actually exists. To do this, we first need some notation. Let us define the set

$$\mathcal{T} := \{\text{trace class, positive, self-adjoint operators on } \mathcal{H}\}.$$

Proposition 4.3.3.1 *Let everything be as before. Then we get that*

$$S_\mu \in \mathcal{T} \iff \int_{\mathcal{H}} \|x\|_{\mathcal{H}}^2 \mathrm{d}\mu(x) < \infty.$$

Proof Assume first that $S_\mu \in \mathcal{T}$ and let $\{e_n\}$ be an orthonormal basis of $\mathcal{H}$. Then we get that

$$\mathrm{Tr}(S_\mu) = \sum_{n=1}^{\infty} \langle S_\mu(e_n), e_n \rangle_{\mathcal{H}} = \sum_{n=1}^{\infty} \int_{\mathcal{H}} \langle x, e_n \rangle_{\mathcal{H}}^2 \mathrm{d}\mu(x) = \int_{\mathcal{H}} \sum_{n=1}^{\infty} \langle x, e_n \rangle_{\mathcal{H}}^2 \mathrm{d}\mu(x) = \int_{\mathcal{H}} \|x\|_{\mathcal{H}}^2 \mathrm{d}\mu(x),$$

by the monotone convergence theorem. Conversely, assume that $\int_{\mathcal{H}} \|x\|_{\mathcal{H}}^2 \mathrm{d}\mu(x) < \infty$. Then, we can define

$$B(x, y) = \int_{\mathcal{H}} \langle x, z \rangle_{\mathcal{H}} \langle y, z \rangle_{\mathcal{H}} \mathrm{d}\mu(z).$$

Hence, we get that

$$|B(x, y)| = \left| \int_{\mathcal{H}} \langle x, z \rangle_{\mathcal{H}} \langle y, z \rangle_{\mathcal{H}} \mathrm{d}\mu(z) \right| \leq \|x\|_{\mathcal{H}} \|y\|_{\mathcal{H}} \int_{\mathcal{H}} \|z\|_{\mathcal{H}}^2 \mathrm{d}\mu(z),$$

and thus B is a bounded bilinear form. Moreover, B is symmetric and positive. Hence, there is a positive self-adjoint operator S_μ such that $B(x, y) = \langle S_\mu(x), y \rangle_{\mathcal{H}}$. Now, we can check that

$$\sum_{n=1}^{\infty} \langle S_\mu(e_n), e_n \rangle_{\mathcal{H}} = \int_{\mathcal{H}} \|x\|_{\mathcal{H}}^2 \mathrm{d}\mu(x) < \infty$$

for any orthonormal basis $\{e_n\}$. Thus, we get that $S_\mu \in \mathcal{T}$. $\qquad\square$

4.3.3.1 Characteristic Functionals

Definition 4.3.3.3 (*Positive-definite function*). A function $\phi: \mathcal{H} \to \mathbb{C}$ is called *positive definite*, if for all $c_1, ..., c_n \in \mathbb{C}$ and $h_1, ..., h_n \in \mathcal{H}$ with $n = 1, 2, ...$ we have

$$\sum_{j,k=1}^{n} c_k \phi(h_k - h_j)\bar{c}_j \geq 0. \tag{4.3.11}$$

Definition 4.3.3.4 (*Characteristic functional III*). Let μ be a Borel measure on $\mathcal{H}$. The *characteristic functional* (or *Fourier transform*) $\hat{\mu}$ of μ is a function $\hat{\mu}: \mathcal{H} \to \mathbb{C}$ defined by

$$\hat{\mu}(y) = \int_{\mathcal{H}} e^{i\langle y, x\rangle_{\mathcal{H}}} d\mu(x). \tag{4.3.12}$$

Remark 4.3.3.2 It is easy to check that:

(1) $|\hat{\mu}(x)| \leq \mu(\mathcal{H})$ for all $x \in \mathcal{H}$.
(2) If μ is a probability measure, then $\hat{\mu}(0) = 1$.
(3) If μ is a finite measure, then $\hat{\mu}$ is uniformly continuous on $\mathcal{H}$.

Lemma 4.3.3.1 *Let μ be a Borel measure on $\mathcal{H}$. Then $\hat{\mu}$ is a positive-definite functional on $\mathcal{H}$.*

Proof Let $h_1, ..., h_n \in \mathcal{H}$ and $c_1, ..., c_n \in \mathbb{C}$. Then we get that

$$\sum_{j,k=1}^{n} c_j \hat{\mu}(h_j - h_k)\bar{c}_k = \int_{\mathcal{H}} \sum_{j,k=1}^{n} c_j e^{i\langle h_j, x\rangle_{\mathcal{H}}} e^{-i\langle h_k, x\rangle_{\mathcal{H}}} \bar{c}_k d\mu(x)$$

$$= \int_{\mathcal{H}} \sum_{j,k=1}^{n} c_j e^{i\langle h_j, x\rangle_{\mathcal{H}}} \overline{e^{i\langle h_k, x\rangle_{\mathcal{H}}} c_k} d\mu(x)$$

$$= \int_{\mathcal{H}} \left| \sum_{j=1}^{n} c_j e^{i\langle h_j, x\rangle_{\mathcal{H}}} \right|^2 d\mu(x) \geq 0.$$

$\square$

Definition 4.3.3.5 (*Gaussian measure III*). A Borel measure μ on $\mathcal{H}$ is called a *Gaussian measure on $\mathcal{H}$*, if for all $h \in \mathcal{H}$ we get that the pushforward measure $(\alpha_h)_* \mu$ is a Gaussian measure on $\mathbb{R}$, where $\alpha_h: \mathcal{H} \to \mathbb{R}$ is defined by $\alpha_h(x) := \langle h, x\rangle$.

Lemma 4.3.3.2 *Let μ be a Gaussian measure on $\mathcal{H}$. Then there are functions m and σ on $\mathcal{H}$ such that*

$$\hat{\mu}(y) = e^{im(y) - \frac{1}{2}\sigma(y)}.$$

Proof Recall at first that $\widehat{(\alpha_h)_* \mu}(t) = \hat{\mu}(th)$. Since the pushforward $(\alpha_h)_* \mu$ is a Gaussian measure, we have that

$$\widehat{(\alpha_h)_*\mu}(t) = e^{im(h)t - \frac{1}{2}t^2\sigma(h)},$$

and thus

$$\hat{\mu}(h) = \widehat{(\alpha_h)_*\mu}(1) = e^{im(h) - \frac{1}{2}\sigma(h)}.$$

$\square$

Exercise 4.3.3.1 Check that for $y \in \mathcal{H}$, we have

$$m(y) = \int_{\mathcal{H}} \langle x, y \rangle_{\mathcal{H}} d\mu(x), \qquad (4.3.13)$$

$$\sigma(y) = \int_{\mathcal{H}} \langle y, x \rangle_{\mathcal{H}}^2 d\mu(x). \qquad (4.3.14)$$

Theorem 4.3.3.1 (Bochner–Kolmogorov–Milnor–Prokhorov). *Let ϕ be a positive-definite functional on $\mathcal{H}$. Then ϕ is a characteristic functional of a Borel probability measure μ on $\mathcal{H}$ if and only if*

(1) $\phi(0) = 0$,
(2) for all $\varepsilon > 0$ there is some $S_\varepsilon \in \mathcal{T}$ such that

$$1 - \text{Re}(\phi(x)) \le \langle S_\varepsilon x, x \rangle_{\mathcal{H}} + \varepsilon, \qquad \forall x \in \mathcal{H}.$$

Proof We refer to (Kuo 1975) for the proof of this theorem. $\square$

Theorem 4.3.3.2 (Prokhorov 1956). *Let everything be as before. Then the following hold:*

(1) $S_\mu \in \mathcal{T}$ whenever μ is a Gaussian measure on $\mathcal{H}$.
(2) $\phi(x) := e^{i\langle m, x \rangle_{\mathcal{H}} - \frac{1}{2}\langle Sx, x \rangle_{\mathcal{H}}}$ is the characteristic functional of a Gaussian measure whenever $m \in \mathcal{H}$ and $S \in \mathcal{T}$.

Proof We will only consider the centered Gaussian measure, i.e., the case of vanishing mean $m(y) = 0$ and thus the case when $\hat{\mu}(x) = e^{-\frac{1}{2}\sigma(x)}$. We want to show that

$$\int_{\mathcal{H}} \|x\|_{\mathcal{H}}^2 d\mu(x) < \infty. \qquad (4.3.15)$$

The idea now is to find some $S \in \mathcal{T}$ and $C_S > 0$ such that

$$\int_{\mathcal{H}} \langle x, y \rangle_{\mathcal{H}}^2 d\mu(y) \le C_S \langle Sx, x \rangle_{\mathcal{H}}, \qquad \forall x \in \mathcal{H}. \qquad (4.3.16)$$

Before we discuss a construction of such an S, let us see why (4.3.16) implies (4.3.15). Consider first an orthonormal basis $\{e_n\}$ of $\mathcal{H}$. Then by (4.3.16), we get that

$$\sum_{n=1}^{\infty} \int_{\mathcal{H}} \langle e_n, y \rangle_{\mathcal{H}}^2 d\mu(y) \le C_S \sum_{n=1}^{\infty} \langle Se_n, e_n \rangle_{\mathcal{H}} = C_S \text{Tr}(S).$$

This implies that

$$\int_{\mathcal{H}} \|y\|_{\mathcal{H}}^2 d\mu(y) = \int_{\mathcal{H}} \sum_{n=1}^{\infty} \langle e_n, y \rangle_{\mathcal{H}}^2 d\mu(y) = \sum_{n=1}^{\infty} \int_{\mathcal{H}} \langle e_n, y \rangle_{\mathcal{H}}^2 d\mu(y) \le C_S \operatorname{Tr}(S) < \infty.$$

Hence, our goal will be to construct S such that (4.3.16) holds. Since μ is a probability measure, we can use Theorem 4.3.3.1 to get that for all $\varepsilon > 0$ there is some $S_\varepsilon \in \mathcal{T}$ such that

$$1 - \hat{\mu}(x) \le \langle S_\varepsilon x, x \rangle_{\mathcal{H}} + \varepsilon, \qquad \forall x \in \mathcal{H}. \tag{4.3.17}$$

Assume now that $\ker(S_\varepsilon) = \{0\}$. In this case we claim that, for all $x \in \mathcal{H} \setminus \{0\}$, we have that

$$\int_{\mathcal{H}} \langle x, y \rangle_{\mathcal{H}}^2 d\mu(x) \le \frac{4}{\varepsilon} \log\left(\frac{1}{1 - 2\varepsilon}\right) \langle S_\varepsilon x, x \rangle_{\mathcal{H}}. \tag{4.3.18}$$

Obviously, (4.3.18) implies (4.3.16). To verify (4.3.18), we proceed as follows: If $y \in \mathcal{H}$ such that $\langle S_\varepsilon y, y \rangle < \varepsilon$, then from (4.3.17) we get $\sigma(y) \le 2 \log\left(\frac{1}{1-2\varepsilon}\right)$. Given $x \in \mathcal{H} \setminus \{0\}$, take $y = \left(\frac{\varepsilon}{2\langle S_\varepsilon x, x \rangle}\right)^{\frac{1}{2}} x$. Then we can check that $\langle S_\varepsilon y, y \rangle < \varepsilon$ and hence we have

$$\sigma(y) \le 2 \log\left(\frac{1}{1 - 2\varepsilon}\right). \tag{4.3.19}$$

Note that we use $\ker(S_\varepsilon) = \{0\}$ to define y. Also, we can check that

$$\sigma(y) = \frac{\varepsilon}{2 \langle S_\varepsilon x, x \rangle_{\mathcal{H}}} \sigma(x).$$

Thus, for all $x \in \mathcal{H} \setminus \{0\}$, we get from (4.3.19) that

$$\sigma(x) \le \frac{4}{\varepsilon} \log\left(\frac{1}{1 - 2\varepsilon}\right) \langle S_\varepsilon x, x \rangle_{\mathcal{H}}.$$

Now using the fact that $\sigma(x) = \int_{\mathcal{H}} \langle x, y \rangle_{\mathcal{H}}^2 d\mu(y)$, we can show (4.3.18) for the case when $\ker(S_\varepsilon) = \{0\}$. If $\ker(S_\varepsilon) = \{0\}$, then we can construct $S \in \mathcal{T}$ with $S_\varepsilon \le S$ and $\ker(S) = \{0\}$ as follows: Let $\{\lambda_n\}$ be positive eigenvalues of S_ε and ϕ_n eigenvectors corresponding to λ_n such that $\|\phi_n\|_{\mathcal{H}} = 1$ and $\phi_n \perp \phi_m$ for $m \ne n$. Moreover, let $\{\psi_j\}$ be an orthonormal basis of $\ker(S_\varepsilon)$. Then $\{\phi_n, \psi_j\}$ forms an orthonormal basis of $\mathcal{H}$. Define the map

$$S: \mathcal{H} \longrightarrow \mathcal{H},$$

$$x \longmapsto S(x) := \sum_n \lambda_n \langle \phi_n, x \rangle_{\mathcal{H}} \phi_n + \sum_j \frac{1}{j^2} \langle \psi_j, x \rangle_{\mathcal{H}} \psi_j.$$

It is then not hard to check that $S \in \mathcal{T}$ and $\ker(S) = \{0\}$, thus we get that (4.3.17) holds if we replace S_ε by S. Hence, repeating the argument above, (4.3.18) holds for S. This completes the proof. $\qquad\qquad\square$

Now let $\mathcal{H}$ be a separable Hilbert space. Let $\mathcal{F}$ be the set of finite rank projections of $\mathcal{H}$, i.e., $p \in \mathcal{F}$ if and only if $p \colon \mathcal{H} \to \mathcal{H}$ is a projection and $\dim p(\mathcal{H}) < \infty$. Let us define the set

$$\mathcal{R} := \{p^{-1}(B) \mid p \in \mathcal{F}, B \subseteq p(\mathcal{H}), B \text{ is Borel measurable}\}.$$

It is then easy to check that $\mathcal{R}$ is an algebra. However, $\mathcal{R}$ is not a sigma algebra, which can be seen as follows. Let $\overline{B(0,1)}$ be the closed unit ball in $\mathcal{H}$ around 0. When $\mathcal{H}$ is infinite-dimensional, $\overline{B(0,1)}$ is not a cylinder set, i.e., $\overline{B(0,1)} \notin \mathcal{R}$, as $C \in \mathcal{R}$ implies that C is unbounded. We claim that $\overline{B(0,1)}$ can be written as countable intersections of elements of $\mathcal{R}$. Let $\{h_n\}$ be a countable dense subset of $\mathcal{H}$ with $h_n \neq 0$ for all n. Moreover, for $N \in \mathbb{N}$, we define the set

$$K_N := \{h \in \mathcal{H} \mid |\langle h, h_n \rangle_\mathcal{H}| \leq \|h_n\|_\mathcal{H}, \forall n = 1, 2, ..., N\}.$$

Exercise 4.3.3.2 Show that $K_N \in \mathcal{R}$ for all $N \in \mathbb{N}$.

It is easy to see that $\overline{B(0,1)} \subseteq \bigcap_{N=1}^{\infty} K_N$. Assume that $h \notin \overline{B(0,1)}$. Then there is some $h' \in \mathcal{H} \setminus \{0\}$ such that $\frac{|\langle h, h' \rangle_\mathcal{H}|}{\|h'\|_\mathcal{H}} \geq \delta + 1$ for some $\delta > 0$. Choose then a subsequence $\{h_{n_k}\}$ such that $h_{n_k} \to h'$. Then $\frac{|\langle h, h_{n_k} \rangle_\mathcal{H}|}{\|h_{n_k}\|_\mathcal{H}} \to \frac{|\langle h, h' \rangle_\mathcal{H}|}{\|h'\|_\mathcal{H}}$ and thus

$$\frac{|\langle h, h_{n_k} \rangle_\mathcal{H}|}{\|h_{n_k}\|_\mathcal{H}} \geq \delta + 1$$

as $k \to \infty$. This shows that $h \notin \bigcap_{N=1}^{\infty} K_N$ and hence we have showed that $\bigcap_{N=1}^{\infty} K_N \subseteq \overline{B(0,1)}$. This means that $\overline{B(0,1)} = \bigcap_{N=1}^{\infty} K_N$. Next, we define a finitely additive measure μ on $\mathcal{R}$ as follows. Let $p \in \mathcal{F}$ and B be a Borel subset of $p(\mathcal{H})$ and $\dim p(\mathcal{H}) = n$. We can then define the measure to be given by

$$\mu(p^{-1}(B)) := \frac{1}{(2\pi)^{\frac{n}{2}}} \int_B e^{-\frac{1}{2}\|x\|_\mathcal{H}^2} dx.$$

Exercise 4.3.3.3 Show that μ is a finitely additive measure on R.

Exercise 4.3.3.4 Show directly that μ cannot be countably additive.

Hence, we cannot hope to construct the standard Gaussian measure on $\mathcal{H}$. Actually, in the infinite-dimensional case the identity operator is not a trace class operator. One can ask whether there is a way to make sense of the standard Gaussian measure on $\mathcal{H}$. The answer is *yes*. There is a way to understand the *standard* Gaussian measure on $\mathcal{H}$. The idea is to *expand* $\mathcal{H}$ so that it supports a countably additive Gaussian measure.

4.3.4 Standard Gaussian Measure on $\mathcal{H}$

The actual question in this case is how we have to *expand* $\mathcal{H}$ such that it works. The technical tool we use here is *Kolmogorov's theorem*. Let us briefly recall this without a proof. Let $\{X_i\}_{i \in \mathcal{I}}$ be a family of topological spaces. Assume that for each finite $I \subseteq \mathcal{I}$, we have a Borel probability measure μ_I on $X_I := \prod_{i \in I} X_i$. Given $J \subseteq I \subseteq \mathcal{I}$, with I finite, let $\pi_{IJ} \colon X_I \to X_J$ denote the projection onto the first J coordinates.

Definition 4.3.4.1 (*Compatible family*). The family $\{X_I, \mu_J\}_{\substack{J \subseteq I \subseteq \mathcal{I}, \\ I \text{ finite}}}$ is said to form a *compatible family*, if for all $J \subseteq I$ we have $(\pi_{IJ})_* \mu_I = \mu_J$.

Theorem 4.3.4.1 (Kolmogorov). *Let* $\{X_I, \mu_J\}_{\substack{J \subseteq I \subseteq \mathcal{I} \\ I \text{ finite}}}$ *be a compatible family. Then there is a unique probability measure $\mu_{\mathcal{I}}$ on $X_{\mathcal{I}} = \prod_{i \in \mathcal{I}} X_i$ and measurable maps $\pi_I \colon X_{\mathcal{I}} \to X_I$ for finite $I \subseteq \mathcal{I}$ such that $(\pi_I)_* \mu_{\mathcal{I}} = \mu_I$.*

To apply Theorem 4.3.4.1 in our situation, we proceed as follows. Let $\{e_n\}$ be an orthonormal basis of $\mathcal{H}$. Define then a measure μ_n on $\mathbb{R}^n$ by

$$\mu_n(B) := \mu(p_n^{-1}(B)),$$

where $p_n \colon \mathcal{H} \to \mathrm{span}\{e_1, ..., e_n\} \cong \mathbb{R}^n$ is the projection and μ the cylindrical measure defined as before. Then it is easy to check that $\{\mathbb{R}^n, \mu_n\}$ forms a compatible family of probability measures. Hence, by Theorem 4.3.4.1 there is a probability space $(\Omega, \tilde{\mu})$ and random variables $\xi_1, ..., \xi_n$ on Ω such that

$$\tilde{\mu}\left(\{\omega \in \Omega \mid (\xi_1(\omega), ..., \xi_n(\omega)) \in B, \ B \subseteq \mathbb{R}^n \text{ Borel measurable}\}\right) \qquad (4.3.20)$$
$$= \mu(\{h \in \mathcal{H} \mid (\langle h, e_1 \rangle_{\mathcal{H}}, ..., \langle h, e_n \rangle_{\mathcal{H}}) \in B\}) \quad (\text{or simply } \mu_n(B)).$$

Lemma 4.3.4.1 *The $\{\xi_i\}$ are independent and identically distributed random variables with mean 0 and variance 1.*

Note that using the ξ_is we can define $\mathcal{H}$-valued random variables X_n by

$$X_n \colon \Omega \longrightarrow \mathcal{H},$$
$$\omega \longmapsto X_n(\omega) := \sum_{i=1}^{n} \xi_i(\omega) e_n. \qquad (4.3.21)$$

Moreover, we get that

$$(p_n \circ X_n)_* \tilde{\mu} = \mu_n. \tag{4.3.22}$$

If $\{X_n\}$ *converges in probability* (i.e., convergence in measure), then it would induce a random variable $X \colon \Omega \to \mathcal{H}$ and hence we would get a measure $X_* \tilde{\mu}$ on $\mathcal{H}$ and by construction it would be the standard Gaussian measure on $\mathcal{H}$. Unfortunately, the bad thing is that the sequence $\{X_n\}$ does not converge in probability. We already know that this is not possible because we have seen that there cannot exist a Gaussian measure μ on $\mathcal{H}$ (assuming $\mathcal{H}$ is infinite-dimensional) whose characteristic functional is $\hat{\mu}(x) = \mathrm{e}^{-\frac{1}{2}\|x\|_{\mathcal{H}}^2}$. Let us see directly how $\{X_n\}$ fails to converge in probability. To show this, it is sufficient to show that $\{X_n\}$ is not a Cauchy sequence in probability.

Lemma 4.3.4.2 *The sequence $\{X_n\}$ is not Cauchy in probability.*

Proof Let $\varepsilon > 0$ and $n > m$. Then

$$\tilde{\mu}\left(\left\{\omega \in \Omega \;\middle|\; \left\|\sum_{i=m+1}^{n} \xi_i(\omega)e_i\right\|_{\mathcal{H}} > \varepsilon\right\}\right) = \mu_{n-m}\left(\mathbb{R}^{n-m} \setminus \overline{B(0,\varepsilon)}\right)$$

$$= 1 - \mu_{n-m}(\overline{B(0,\varepsilon)}) \tag{4.3.23}$$

$$\geq 1 - \mu_{n-m}([-\varepsilon,\varepsilon]^{n-m})$$

$$= 1 - (\mu_1([-\varepsilon,\varepsilon]))^{n-m}.$$

Note that $\mu_1([-\varepsilon,\varepsilon]) < 1$ implies that $(\mu_1([-\varepsilon,\varepsilon]))^{n-m} \xrightarrow{n,m \to \infty} 0$. Here, we have that $\mu_1([-\varepsilon,\varepsilon]) = \frac{1}{\sqrt{2\pi}} \int_{-\varepsilon}^{\varepsilon} \mathrm{e}^{-\frac{1}{2}x^2} \mathrm{d}x$. This implies that $\{X_n\}$ is not Cauchy in probability. $\qquad\square$

Hence, our strategy to construct a Banach space containing $\mathcal{H}$ would be the following. First consider a new norm $\|\cdot\|_W$ for which the sequence $\{X_n\}$ is Cauchy in probability. In this case $\{X_n\}$ converges in probability if we consider the Banach space obtained by completing $\mathcal{H}$ with respect to this new norm. This motivates the following definition.

Definition 4.3.4.2 (*Measurable norm*). A norm $\|\cdot\|_W$ on $\mathcal{H}$ is said to be *measurable* if for all $\varepsilon > 0$ there is some $p_0 \in \mathcal{F}$ such that

$$\mu(\{h \in \mathcal{H} \mid \|p_h\|_W > \varepsilon\}) < \varepsilon$$

for all $p \in \mathcal{F}$ such that $p \perp p_0$.

Geometrically, it means that $\|\cdot\|_W$ is such that μ is concentrated in a tubular neighborhood of some $p_0 \in \mathcal{F}$. A non-example would be the norm $\|\cdot\|_{\mathcal{H}}$ on $\mathcal{H}$, which is not measurable.

Theorem 4.3.4.2 (*Gross*). *Let $\|\cdot\|_W$ be a measurable norm on $\mathcal{H}$. Let W be the Banach space obtained by the completion of $\mathcal{H}$ with respect to $\|\cdot\|_W$. Then the sequence $\{X_n\}$ converges in probability in W.*

Theorem 4.3.4.3 (Gross). *Given a separable real Hilbert space $\mathcal{H}$, there is a separable Banach space W with a linear continuous dense embedding $\iota: \mathcal{H} \hookrightarrow W$ and a Gaussian measure μ_W on W such that*

$$\mu_W(\{w \in W \mid (f_1(w), ..., f_n(w)) \in B, B \subseteq \mathbb{R}^n \text{ Borel measurable}\}) \quad (4.3.24)$$
$$= \mu(\{h \in \mathcal{H} \mid (\langle h, f_1 \rangle_{\mathcal{H}}, ..., \langle h, f_n \rangle_{\mathcal{H}}) \in B\})$$

for all $f_1, ..., f_n \in W^* \hookrightarrow \mathcal{H}^* \cong \mathcal{H}$. In particular, for all $f \in W^* \subseteq \mathcal{H}$, we get that

$$\hat{\mu}(f) = e^{-\frac{1}{2}\|f\|_{\mathcal{H}}^2}.$$

Here, Gaussian measure means that for all $f \in W^*$ we have that $f_*\mu$ is Gaussian on $\mathbb{R}$.

Remark 4.3.4.1 If $\| \cdot \|_W$ is a measurable norm on $\mathcal{H}$, then there is some $c > 0$ such that $\|h\|_W \le c\|h\|_{\mathcal{H}}$ for all $h \in \mathcal{H}$ (cf. Kuo (1975)). It was expected that $\| \cdot \|_W$ is dominated by $\| \cdot \|_{\mathcal{H}}$ because we needed a bigger topology on $\mathcal{H}$ to allow convergence of $\{X_n\}$.

Remark 4.3.4.2 Let A be a positive Hilbert–Schmidt operator on $\mathcal{H}$. Define a new norm by

$$\|h\|_{W_A} = \|Ah\|_{\mathcal{H}}.$$

Then $\| \cdot \|_{W_A}$ is actually a measurable norm on $\mathcal{H}$ (cf. Kuo (1975)).

Remark 4.3.4.3 By Remark 4.3.4.2 we see that there can be many such Banach spaces. In other words, we have no uniqueness. However, we actually do not care about that.

To elaborate on Remark 4.3.4.3, we want to emphasize that $\mathcal{H}$ contains all information about the measure. The next goal is to make this statement a little more precise, and this requires some effort. Given a separable Hilbert space $\mathcal{H}$, we have seen that there is a Banach space W, a linear continuous dense embedding $\iota: \mathcal{H} \hookrightarrow W$ and a Gaussian measure μ on W such that $\hat{\mu}(f) = e^{-\frac{1}{2}\|f\|_{\mathcal{H}}^2}$, where $f \in W^* \subseteq \mathcal{H}^* \cong \mathcal{H}$. Next, we would like to understand whether it is possible to identify $\mathcal{H}$ from a separable Banach space W and a centered Gaussian measure μ. More precisely, given a separable Banach space W and a centered Gaussian measure μ on W, is it possible to find a Hilbert space $\mathcal{H}(\mu)$ together with a linear continuous dense embedding $\iota: \mathcal{H}(\mu) \hookrightarrow W$ such that $\hat{\mu}(f) = e^{-\frac{1}{2}\|f\|_{\mathcal{H}(\mu)}^2}$? We will start with a separable real Hilbert space $\mathcal{H}$, a Banach space W and a Gaussian measure μ given by Theorem 4.3.4.3. Then we want to understand how we can recover $\mathcal{H}$ from W and μ. This will give a hint on the construction of $\mathcal{H}(\mu)$ out of W and μ. Let $\mathcal{H}$, W and μ be as in Theorem 4.3.4.2. Given $f \in W^*$, we define $q_\mu(f) := \int_W f(w)^2 d\mu(w)$. More generally, we can define

$$q_\mu: W^* \times W^* \longrightarrow \mathbb{R},$$

$$(f, g) \longmapsto q_\mu(f, g) = \int_W f g \mathrm{d}\mu.$$

Definition 4.3.4.3 (*Covariance of a measure*). The map q_μ is called the *covariance* of μ.

Exercise 4.3.4.1 Show that $q_\mu(f, g) = \langle f, g \rangle_{\mathcal{H}}$, where $f, g \in W^* \subseteq \mathcal{H}^* \cong \mathcal{H}$.

First, we would like to show that q_μ is a continuous positive-definite symmetric bilinear form on W^*. To see this, we need a technical tool which is called *Fernique's theorem*. We will state it without a proof.

Theorem 4.3.4.4 (Fernique 1970). *Let W be a separable Banach space and μ a Gaussian measure on W. Then there is some $\varepsilon = \varepsilon(\mu) > 0$ such that*

$$\int_W \mathrm{e}^{\varepsilon \|w\|_W^2} \mathrm{d}\mu(w) < \infty.$$

Corollary 4.3.4.1 *Let everything be as before. Then we get that*

$$\int_W \|w\|_W^p \mathrm{d}\mu(w) < \infty, \quad \forall p \geq 1.$$

Proposition 4.3.4.1 *The covariance map q_μ is a continuous bilinear form on W^*.*

Proof Note that, for all $f, g \in W^*$, we have

$$|q_\mu(f, g)| \leq \int_W |f(w) g(w)| \mathrm{d}\mu(w) \leq \|f\|_{W^*} \|g\|_{W^*} \underbrace{\int_W \|w\|_W^2 \mathrm{d}\mu(w)}_{\text{constant}}.$$

$\square$

Note that $f \in W^*$ implies that $q_\mu(f) < \infty$ and thus $f \in L^2(W, \mu)$. Hence, we have a canonical linear map

$$T : W^* \longrightarrow L^2(W, \mu),$$
$$f \longmapsto T(f).$$

Lemma 4.3.4.3 *The map T is continuous.*

Proof Note that we have

$$\|T(f)\|_{L^2(W,\mu)} = \int_W f(w)^2 \mathrm{d}\mu(w) \leq \|f\|_{W^*}^2 \int_W \|w\|_W^2 \mathrm{d}\mu(w).$$

$\square$

Corollary 4.3.4.2 *The norm on W^* induced by q_μ is weaker then $\|\cdot\|_{W^*}$.*

Lemma 4.3.4.4 *Let J be the map $J\colon W^* \to \mathcal{H}$ given as the composition $W^* \hookrightarrow \mathcal{H}^* \xrightarrow{\sim} \mathcal{H}$. Then $J\colon (W^*, q_\mu) \to \mathcal{H}$ is a linear continuous dense isometric embedding.*

Proof The proof is a direct consequence of Corollary 4.3.4.2. We will leave the details to the reader. $\qquad\square$

Exercise 4.3.4.2 Given $h \in \mathcal{H}$, define $\alpha_h\colon W^* \to \mathbb{R}$ by $\alpha_h(f) := f(h)$. Show that α_h is continuous on W^* with respect to the topology induced by q_μ.

Corollary 4.3.4.3 *If $h \in \mathcal{H}$, then $\|h\|_{\mathcal{H}} = \sup\limits_{f \in W^*\backslash\{0\}} \dfrac{|f(h)|}{\sqrt{q_\mu(f,f)}}$.*

Proof Since $J(W^*)$ is dense in $\mathcal{H}$, we know that

$$\|h\|_{\mathcal{H}} = \sup_{f \in W^*\backslash\{0\}} \frac{|f(h)|}{\|J(f)\|} = \sup_{f \in W^*\backslash\{0\}} \frac{|f(h)|}{\|f\|_{L^2(W,\mu)}} = \sup_{f \in W^*\backslash\{0\}} \frac{|f(h)|}{\sqrt{q_\mu(f,f)}}.$$

$\qquad\square$

Let K be the completion of $T(W^*)$ in $L^2(W, \mu)$. Then we can observe that J extends to an isometry $J\colon K \to \mathcal{H}$.

Exercise 4.3.4.3 Show that $J\colon K \to \mathcal{H}$ is an isomorphism of Hilbert spaces.

Let us briefly summarize what we have seen so far:

(1) We have seen that

$$\|h\|_{\mathcal{H}} = \sup_{f \in W^*\backslash\{0\}} \frac{|f(h)|}{\sqrt{q_\mu(f,f)}}, \qquad \forall h \in \mathcal{H}.$$

This relation can be thought of as constructing the norm on $\mathcal{H}$ out of W and μ. It will be the key to constructing $\mathcal{H}$ out of W and μ.
(2) The map $J\colon K \to \mathcal{H}$ is an isomorphism of Hilbert spaces. In particular, it is an isomorphism of Banach spaces.

Given a separable Banach space on W and a centered Gaussian measure μ on W, point (1) will be used to define a normed space $\mathcal{H}(\mu)$ and point (2) will be used to give an inner product on $\mathcal{H}(\mu)$. This way we will be able to construct $\mathcal{H}(\mu)$ out of W and μ.

Definition 4.3.4.4 ($\mathcal{H}(\mu)$-norm). Let W be a real separable Banach space and μ a centered Gaussian measure on W. Define a *norm* $\|\cdot\|_{\mathcal{H}(\mu)}$ on W by

$$\|w\|_{\mathcal{H}(\mu)} := \sup_{f \in W^*\backslash\{0\}} \frac{|f(w)|}{\sqrt{q_\mu(f,f)}}.$$

Definition 4.3.4.5 (*Cameron–Martin space* (Cameron and Martin 1944)). The *Cameron–Martin space* is defined as

$$\mathcal{H}(\mu) := \{w \in W \mid \|w\|_{\mathcal{H}(\mu)} < \infty\}.$$

Exercise 4.3.4.4 Show that $\mathcal{H}(\mu)$ is a normed space.

Exercise 4.3.4.5 Show that $w \in \mathcal{H}(\mu)$ if and only if the map $f \mapsto f(w)$ is continuous on W^* if W^* has the topology induced by q_μ.

Proposition 4.3.4.2 *The Cameron–Martin space $\mathcal{H}(\mu)$ is a Banach space, i.e., it is complete with respect to the norm $\| \cdot \|_{\mathcal{H}(\mu)}$.*

Proof We first show that there is some constant $c > 0$ such that for all $w \in \mathcal{H}(\mu)$ we get that

$$\|w\|_W \leq c\|w\|_{\mathcal{H}(\mu)}.$$

In other words, there is a continuous embedding $\iota \colon (\mathcal{H}(\mu), \| \cdot \|_{\mathcal{H}(\mu)}) \hookrightarrow W$. Let $w \in W \setminus \{0\}$. By the *Hahn–Banach theorem* we can choose $f \in W^*$ such that $\|f\|_{W^*} = 1$ and $f(w) = \|w\|_W$. Moreover, by Proposition 4.3.4.1 we have that $\|f\|_{q_\mu} \leq \tilde{c}\|f\|_{W^*}$ and thus $\|f\|_{q_\mu} \leq \tilde{c}$ for some constant $\tilde{c} > 0$. This means that we have

$$\|w\|_W = f(w) = |f(w)| = \frac{|f(w)|}{\|f\|_{W^*}} \leq c\frac{|f(w)|}{\|f\|_{q_\mu}} \leq c\|w\|_{\mathcal{H}(\mu)},$$

with $c := \frac{1}{\tilde{c}}$. To show that $\mathcal{H}(\mu)$ is complete, let $\{h_n\}$ be a Cauchy sequence in $\mathcal{H}(\mu)$ with respect to the norm $\| \cdot \|_{\mathcal{H}(\mu)}$. By the previous section it is also Cauchy in $(W, \| \cdot \|_W)$. Since W is complete, there is some $h \in W$ such that $h_n \xrightarrow{n \to \infty} h$ in $(W, \| \cdot \|_W)$. We thus claim that $h \in \mathcal{H}(\mu)$ and $h_n \xrightarrow{n \to \infty} h$ in $\mathcal{H}(\mu)$. Let $\varepsilon > 0$ and choose $m \geq n$ large enough such that

$$f(h_n - h) < \varepsilon\sqrt{q_\mu(f, f)}.$$

In order to do this, we use that $h_n \xrightarrow{n \to \infty} h$ in W and that $f \in W^*$. Therefore, we get that

$$\frac{|f(h_n - h)|}{\sqrt{q_\mu(f, f)}} \leq \frac{|f(h_n - h_m)|}{\sqrt{q_\mu(f, f)}} + \frac{|f(h_m - h)|}{\sqrt{q_\mu(f, f)}} \leq \|h_n - h_m\|_{\mathcal{H}(\mu)} + \varepsilon.$$

This shows that $\|h_n - h\|_{\mathcal{H}(\mu)} < \infty$ implies that $h \in \mathcal{H}(\mu)$ and $\|h_n - h\|_{\mathcal{H}(\mu)} \xrightarrow{n \to \infty} 0$. $\qquad\square$

Remark 4.3.4.4 Even though we have not been able to show that $\mathcal{H}(\mu)$ is a Banach space, so far, we have not done anything to show that $\mathcal{H}(\mu) \neq \{0\}$. In order to see this, and that $\mathcal{H}(\mu)$ is a Hilbert space, we will use the notion of *Bochner integrals*.

4.3.4.1 Digression on Bochner Integrals

Let $(\Omega, \sigma(\Omega), \mu)$ be a measure space and W a Banach space.

Definition 4.3.4.6 (*Bochner integrable*). Let $f : \Omega \to W$ be a measurable map, where we consider the Borel sigma algebra on W. We say f is *Bochner integrable*, if

$$\int_\Omega \|f(\omega)\|_W d\mu(\omega) < \infty.$$

If f is Bochner integrable, it is possible to define $\int_\Omega f(\omega) d\mu(\omega) \in W$, i.e., a W-valued integral on Ω, which is called the *Bochner integral* of f. For us $(\Omega, \sigma(\Omega), \mu)$ will be $(W, \mathcal{B}(W), \mu)$. We can see that for $f \in W^*$ we get that

$$\int_W \|wf(w)\|_W d\mu(w) \le \|f\|_{W^*} \int_W \|w\|_W^2 d\mu(w) < \infty,$$

by Fernique's theorem (Theorem 4.3.4.4). This implies that, given $f \in W^*$, the map $w \mapsto wf(w)$ is Bochner integrable. Hence, we get that $\int_W wf(w) d\mu(w) \in W$. Define a map $J : W^* \to W$ by

$$J(f) := \int_W wf(w) d\mu(w) \in W.$$

Exercise 4.3.4.6 Show that $J(f) \in \mathcal{H}(\mu)$ and $\|J(f)\|_{\mathcal{H}(\mu)} \le \|f\|_{q_\mu}$.

As a consequence of Exercise 4.3.4.6, we see that $J : W^* \to \mathcal{H}(\mu)$ is a contraction if W^* is endowed with the norm induced by q_μ, i.e., we have

$$\|f\|_{q_\mu} = \|f\|_{L^2(W, \mu)}.$$

This shows that $\mathcal{H}(\mu) \ne \{0\}$. Since W^* is dense in K, we have an isometry $J : K \to \mathcal{H}(\mu)$. Next, we want to show that J is surjective. Let $h \in \mathcal{H}(\mu)$ such that the map $f \mapsto f(h)$ is continuous on (W^*, q_μ). We will call this map $\hat{h}$. Note that $\hat{h}$ extends to a continuous linear functional on K. Hence, by *Riesz's representation theorem*, $\hat{h}$ can be identified with an element of $h \in K$. It is easy to check that $J(\hat{h}) = h$. Thus J is an isomorphism of Banach spaces. Now we can endow $\mathcal{H}(\mu)$ with the Hilbert space structure induced by J. Since K is a separable Hilbert space, so is $\mathcal{H}(\mu)$. Moreover, we have the following theorem.

Theorem 4.3.4.5 (Gross). *For any real separable Banach space W and Gaussian measure μ, there exist a Hilbert space $\mathcal{H}(\mu) \subseteq W$ such that $\mathcal{H}(\mu) \hookrightarrow W$ is a linear continuous dense embedding and*

$$\hat{\mu}(f) = e^{-\frac{1}{2}\|f\|_{\mathcal{H}(\mu)}^2}, \quad \forall f \in W^*.$$

Exercise 4.3.4.7 Let $\mathcal{H}(\mu)$ be the Cameron–Martin space of (W, μ) and let $\{e_n\} \subseteq W^*$ be an orthonormal basis of $\mathcal{H}(\mu)$. Show that for any $w \in W$, we have that $w \in \mathcal{H}(\mu)$ if and only if

$$\sum_{n=1}^{\infty} e_n(w) < \infty,$$

and in this case we have

$$\|w\|_{\mathcal{H}(\mu)}^2 = \sum_{n=1}^{\infty} e_n(w)^2.$$

Corollary 4.3.4.4 *If $\mathcal{H}(\mu)$ is infinite-dimensional, we get that $\mu(\mathcal{H}(\mu)) = 0$.*

Proof Let everything be as in Exercise 4.3.4.7. Then $\{e_n\}$ is a sequence of independent, identically distributed random variables with mean 0 and variance 1. Hence, by the law of large numbers, we get that

$$\sum_{n=1}^{\infty} e_n(w)^2 = \mu \quad \text{almost everywhere.}$$

However, by Exercise 4.3.4.7 we have

$$\mathcal{H}(\mu) = \left\{ w \in W \,\middle|\, \sum_{n=1}^{\infty} e_n(w)^2 < \infty \right\}.$$

This implies that $\mu(\mathcal{H}(\mu)) = 0$. $\square$

There is actually another way of interpreting the Cameron–Martin space.

Theorem 4.3.4.6 (Cameron and Martin 1944). *Let $\mathcal{H}(\mu)$ be the Cameron–Martin space of (W, μ), $h \in W$ and $T_h \colon W \to W$ given by $T_h(w) = w - h$. If $h \in \mathcal{H}(\mu)$, then $\mu_h = (T_h)_* \mu$ is absolutely continuous with respect to μ (i.e., we have $\mu_h \ll \mu$) and the Radon–Nikodym derivative is given by*

$$\frac{\mathrm{d}\mu_h}{\mathrm{d}\mu} = \mathrm{e}^{-\frac{1}{2}\|h\|_{\mathcal{H}(\mu)}^2} \mathrm{e}^{\langle h, \cdot \rangle}. \tag{4.3.25}$$

Proof In order to prove the theorem, we will compute the Fourier transform of μ_h and $\mathrm{e}^{-\frac{1}{2}\|h\|_{\mathcal{H}(\mu)}^2} \mathrm{e}^{\langle h, \cdot \rangle} \mu$. Thus, consider $f \in W^*$ and note that we then have

$$\hat{\mu}_h(f) = \int_W \mathrm{e}^{\mathrm{i}f(w)} \mathrm{d}\mu_h(w) = \int_W \mathrm{e}^{\mathrm{i}f(w-h)} \mathrm{d}\mu(w) = \mathrm{e}^{-\mathrm{i}f(h)} \mathrm{e}^{-\frac{1}{2}q_\mu(f,f)}.$$

On the other hand, defining $\tilde{h} := J^{-1}(h)$, we get that

$$
\begin{aligned}
\int_W e^{-\frac{1}{2}\|h\|^2} e^{\tilde{h}(w)} e^{\mathrm{i}f(w)} \mathrm{d}\mu(w) &= e^{-\frac{1}{2}\|h\|^2} \int_W e^{\mathrm{i}(f-\mathrm{i}\tilde{h})(w)} \mathrm{d}\mu(w) \\
&= e^{-\frac{1}{2}\|h\|^2} e^{-\frac{1}{2}q_\mu(f-\mathrm{i}h,f-\mathrm{i}h)} \\
&= e^{\mathrm{i}f(h)} e^{-\frac{1}{2}q_\mu(f,f)}.
\end{aligned}
\tag{4.3.26}
$$

Hence, we get that $(T_h)_*\mu$ is absolutely continuous with respect to μ and therefore (4.3.25) holds. $\qquad\square$

Remark 4.3.4.5 We get that μ and μ_h are mutually singular whenever $h \in W \setminus \mathcal{H}(\mu)$. Hence, we get that $h \in \mathcal{H}(\mu)$ if and only if $(T_h)_*\mu \ll \mu$.

Example 4.3.4.1 (Cameron–Martin space of the classical Wiener space). Recall the classical Wiener space which is given as $W = C_0([0, 1])$ endowed with the Wiener measure μ together with the covariance

$$
q_\mu(\mathrm{ev}_t, \mathrm{ev}_s) = \min\{s, t\},
$$

where ev denotes the evaluation map. One can check that $E = \mathrm{span}\{\mathrm{ev}_t \mid t \in [0, 1]\}$ is dense in W^*. First, let us compute $J: E \to W$. By definition, we have

$$
(J(\mathrm{ev}_t))(s) = \int_{C_0([0,1])} x(s)\, \mathrm{ev}_t(x) \mathrm{d}\mu(x) = \int_{C_0([0,1])} x(s)x(t) \mathrm{d}\mu(x) = \min\{s, t\},
$$

and thus $\frac{\mathrm{d}}{\mathrm{d}s}(J(\mathrm{ev}_t)) = \chi_{[0,t]}$. Note that $q_\mu(\mathrm{ev}_t, \mathrm{ev}_s) = \min\{s, t\}$. On the other hand, we get that

$$
\int_0^1 \dot{J}(\mathrm{ev}_t)(u)\dot{J}(\mathrm{ev}_s)(u)\mathrm{d}u = \int_0^1 \chi_{[0,1]}(u)\chi_{[0,s]}(u)\mathrm{d}u = \min\{s, t\}.
$$

This shows that if $x, y \in \mathcal{H}(\mu)$, we get that

$$
\langle x, y\rangle_{\mathcal{H}(\mu)} = \int_0^1 \dot{x}(u)\dot{y}(u)\mathrm{d}u.
\tag{4.3.27}
$$

Thus, we can write down the Cameron–Martin space $\mathcal{H}(\mu) = \{x \in C_0([0, 1]) \mid \dot{x} \in L^2([0, 1])\}$ and the inner product on $\mathcal{H}(\mu)$ is given by (4.3.27). This means that the Wiener measure is the standard Gaussian measure on the space of 1-Sobolev paths $H^1([0, 1])$. This observation is due to Cameron–Martin, and was later generalized by Gross.

4.4   Wick Ordering

4.4.1   Motivating Example and Construction

Let $H_n(x)$ be the *Hermite polynomial* of degree n on $\mathbb{R}$. It can be defined recursively as follows:

$$H_0(x) := 1,$$
$$H_{n+1}(x) := x H_n(x) - H_{n-1}(x), \quad n \geq 1,$$
$$\frac{\mathrm{d}}{\mathrm{d}x} H_n(x) := n H_{n-1}(x), \quad n \geq 1,$$
$$\int_{\mathbb{R}} H_n(x) H_m(x) \mathrm{d}\mu(x) := 0, \quad \text{for all } n \neq m, \text{ where } \mu \text{ is the standard Gaussian measure on } \mathbb{R}.$$

Moreover, Hermite polynomials are given by the *generating function*

$$e^{tx - \frac{1}{2}t^2} = \sum_{n=0}^{\infty} \frac{t^n}{n!} H_n(x), \tag{4.4.1}$$

more precisely, this means that $e^{tx - \frac{1}{2}t^2}$ is the generating function for $H_n(x)$. This is usually considered as one of the best tools to study the properties of $H_n(x)$.

Exercise 4.4.1.1 Show that $H_n(x) = e^{-\frac{\Delta}{2}}(x^n)$, where $\Delta = -\frac{\mathrm{d}^2}{\mathrm{d}x^2}$.[5]

The following statements hold:

(1) The polynomials $H_n(x)$ form an orthonormal basis of $L^2(\mathbb{R}, \mu)$.
(2) We get that

$$\int_{\mathbb{R}} H_n(x)^2 \mathrm{d}\mu(x) = n!.$$

Note that here $L^2(\mathbb{R}, \mu) := \bigoplus_{n=0}^{\infty} \mathrm{span}(H_n(x))$, i.e., we have $H_n(x) \perp \bigoplus_{k=0}^{n-1} \mathrm{span}(H_k(x))$.
More generally, we get that

$$H_n(x^{t_1}, ..., x^{t_r}) = H_{t_1}(x) \cdots H_{t_r}(x), \quad \text{for } t_1 + \cdots + t_r = n,$$

and one can check that the Hermite polynomials form an orthogonal basis of $L^2(\mathbb{R}^n, \mu)$. On the other hand, $\mathbb{R}$ is a Hilbert space, in fact it is also a Cameron–Martin space for the standard Gaussian measure μ. We can then talk about the *Bosonic Fock space* of $\mathbb{R}$. In particular, we have

---

[5] This can be used as a conceptual way to think about *Wick ordering*.

$$\widetilde{\mathrm{Sym}}^{\bullet}(\mathbb{R}) \;=\!=\!=\; L^2(\mathbb{R}, \mu)$$
$$\| \qquad\qquad \|$$
$$\bigoplus_{n\geq0} \mathrm{Sym}^n(\mathbb{R}) \;\xrightarrow{\sim}\; \bigoplus_{n\geq0} \mathrm{span}(H_n(x))$$

where the arrow on the bottom represents a canonical isomorphism. More generally, we want to prove that

$$L^2(W, \mu) \cong \widetilde{\mathrm{Sym}}^{\bullet}(H(\mu)),$$

where the isomorphism is canonical. Thus, let W be a separable Banach space and μ a centered Gaussian measure on W and $H(\mu)$ be its Cameron–Martin space with covariance q_μ. Let $f \in W^*$ (or $f \in H(\mu)$, since it does not really matter). Define then recursively a sequence of functions $: f^n :$ given by[6]

$$: f^0 := 1,$$
$$\frac{\partial}{\partial f} : f^n := n : f^{n-1} :, \quad n \geq 1,$$
$$\int_W : f^n : \mathrm{d}\mu = 0, \quad n \geq 1.$$

This definition says that $: f^n :$ comes from f^n in the same way as $H_n(x)$ comes from x^n. Let us use generating functions to observe properties of $: f^n :$. Let us define then

$$: e^{\alpha f} :\, := \sum_{k=0}^{\infty} \frac{\alpha^k}{k!} : f^k : .$$

One can then check that

$$\int_W : e^{\alpha f} : \mathrm{d}\mu = 1. \qquad (4.4.2)$$

Moreover, by definition, we get that

$$\frac{\partial}{\partial f} : e^{\alpha f} := \alpha : e^{\alpha f} :,$$

which implies that $: e^{\alpha f} := c \cdot e^{\alpha f}$, for some constant c, which needs to be determined. Using (4.4.2), it is easy to show that $c = \frac{1}{\int_W e^{\alpha f} \mathrm{d}\mu}$.

Exercise 4.4.1.2 Use Wick's theorem to show that

$$\int_W f^{2n+1} \mathrm{d}\mu = 0, \quad \forall n \geq 0,$$

---

[6] The double dots $:\,:$ are the usual way to denote the *normal ordering (Wick ordering) operator* in the physics literature.

$$\int_W f^{2n}\mathrm{d}\mu = \frac{(2n)!}{2^n n!}q_\mu(f,f)^n, \quad \forall n \geq 0.$$

Hence, regarding $\mathrm{e}^{\alpha f}$ as a formal power series in α, we see that

$$\int_W \mathrm{e}^{\alpha f}\mathrm{d}\mu = \sum_{k=0}^\infty \frac{\alpha^k}{k!}\int_W f^k \mathrm{d}\mu = \sum_{k=0}^\infty \frac{\alpha^{2k}}{(2k)!}\frac{(2k)!}{k!}(q_\mu(f,f))^k = \mathrm{e}^{\frac{1}{2}\alpha^2 q_\mu(f,f)},$$

because $:\mathrm{e}^{\alpha f}: = \mathrm{e}^{\alpha f}\mathrm{e}^{-\frac{1}{2}\alpha^2 q_\mu(f,f)}$, which is exactly (4.4.1) when $W = \mathbb{R}$ and $f(x) = x$. One can use the generating function for $:f^n:$ to show that

$$:f^n: = \sum_{k=0}^{\lfloor \frac{n}{2}\rfloor} \frac{n!}{k!(n-2k)!}f^{n-2k}\left(-\frac{1}{2}q_\mu(f,f)\right)^k,$$

$$f^n = \sum_{k=0}^{\lfloor \frac{n}{2}\rfloor} \frac{n!}{k!(n-2k)!}:f^{n-2k}:\left(\frac{1}{2}q_\mu(f,f)\right)^k.$$

Proposition 4.4.1.1 *Let everything be as before. Then we get that*

$$\int_W :f^n::g^m:\mathrm{d}\mu = \delta_{nm}n!\langle f,g\rangle^n_{W^*}, \quad \forall n \neq m, \forall f,g \in W^*.$$

Proof Note that we have

$$:\mathrm{e}^{\alpha f}: = \mathrm{e}^{\alpha f - \frac{1}{2}\alpha^2 q_\mu(f,f)},$$

$$:\mathrm{e}^{\beta g}: = \mathrm{e}^{\beta g - \frac{1}{2}\beta^2 q_\mu(g,g)}.$$

Thus, we get that

$$:\mathrm{e}^{\alpha f}::\mathrm{e}^{\beta g}: =: \mathrm{e}^{\alpha f+\beta g}:\mathrm{e}^{\alpha\beta q_\mu(f,g)}.$$

Hence, we have

$$\int_W :\mathrm{e}^{\alpha f}::\mathrm{e}^{\beta g}:\mathrm{d}\mu = \mathrm{e}^{\alpha\beta q_\mu(f,g)}\underbrace{\int_W :\mathrm{e}^{\alpha f+\beta g}:\mathrm{d}\mu}_{=1} = \sum_{k=0}^\infty \frac{(\alpha\beta)^k}{k!}q_\mu(f,g)^k.$$

Finally, comparing the coefficients of $(\alpha\beta)^k$ we get

$$\frac{1}{(k!)^2}\int_W :f^k::g^k:\mathrm{d}\mu = \frac{1}{k!}q_\mu(f,g)^k.$$

$\square$

The next natural thing is to understand how we can define more generally : $f^{k_1} g^{k_2}$:. The idea is to use the recursive definition, similarly as before. Therefore, for $n = n_1 + \cdots + n_k$, we get

$$: f_1^0 \cdots f_k^0 := 1,$$

$$\int_W : f_1^{n_1} \cdots f_k^{n_k} : d\mu = 0,$$

$$\frac{\partial}{\partial f_i} : f_1^{n_1} \cdots f_k^{n_k} := n_i : f_1^{n_1} \cdots f_i^{n_i-1} \cdots f_k^{n_k} :.$$

Moreover, one can easily check that : $(\alpha f + \beta g)^n := \sum_{k=0}^{n} \binom{n}{k} \alpha^k \beta^{n-k} : f^k g^{n-k} :$ by using generating functions.

Exercise 4.4.1.3 Show that

$$\int_W : f_1 - f_n :: g_1 - g_m : d\mu = \begin{cases} 0, & \text{if } m \neq n, \\ \sum_{\sigma \in S_n} \prod_{i=1}^{n} \langle f_i, g_{\sigma(i)} \rangle_{W^*}, & \text{if } m = n. \end{cases}$$

As in the finite-dimensional case, given a polynomial function P on W, i.e., a function of the form

$$P(w) = \sum_{i_1, \ldots, i_k} a_{i_1} \cdots a_{i_k} f_{i_1}^{\alpha_1}(w) \cdots f_{i_k}^{\alpha_k}(w),$$

where $a_{i_1}, \ldots, a_{i_k} \in \mathbb{R}$ and $f_{i_1}, \ldots, f_{i_k} \in W^*$. We would like to find some operator Δ such that we can formally write

$$: P := e^{-\frac{\Delta}{2}} P,$$

One can use the Cameron–Martin space to make sense of Δ. Take an orthonormal basis $\{e_n\}$ of $\mathcal{H}(\mu)$. Then, loosely speaking, we can define $\Delta := -\sum_{n=1}^{\infty} \frac{\partial^2}{\partial e_n^2}$.

4.4.2 Wick Ordering as a Value of Feynman Diagrams

Given $f_1, f_2, f_3, f_4 \in W^*$, and $\gamma \in \{(1, 2), (3, 4)\}$, we can construct a diagram as follows:

$$f_1 \bullet \!\!\xrightarrow{\quad q_\mu \quad}\!\! \bullet f_2$$

$$f_3 \bullet \!\!\xrightarrow{\quad q_\mu \quad}\!\! \bullet f_4$$

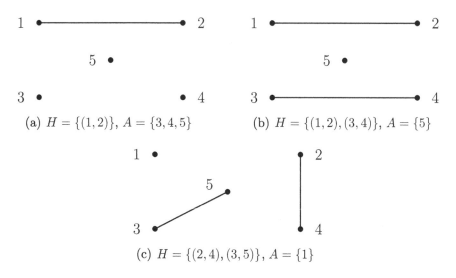

Fig. 4.1 Examples of Feynman graphs with $V = \{1, 2, 3, 4, 5\}$ and different H and A

The value of such a diagram will be the product $q_\mu(f_1, f_2)q_\mu(f_3, f_4)$ (Fig. 4.1).

Definition 4.4.2.1 (*Feynman diagram*). A *Feynman diagram* with n vertices and rank r, where $r \leq \frac{n}{2}$, consists of a set V called the *set of vertices* (thus $|V| = n$), and a set H called the *set of half edges*, which consists of r disjoint pairs of vertices. The remaining vertices are called *unpaired vertices*, which will be denoted by A.

Example 4.4.2.1 If we are given $f_1, ..., f_n \in W^*$, we can think of them as vertices of a Feynman diagram. Hence, given $f_1, ..., f_n \in W^*$ and a Feynman diagram of rank r, i.e., a graph of the form

$$\gamma(H) = \{(i_1, j_1), ..., (i_n, j_n)\},$$

we define the value $F(f_1 \cdots f_n; \gamma)$ of the Feynman diagram as

$$F(f_1 \cdots f_n; \gamma) := \prod_{k=1}^{r} q_\mu(f_{i_k}, f_{j_k}) \prod_{i \in A} f_i.$$

Moreover, we say that γ is *complete* if $n = 2r$. With this notation, we can rephrase Wick's theorem as

$$\int_W f_1 \cdots f_n d\mu = \sum_{\substack{\gamma \text{ complete} \\ \text{Feynman diagram}}} F(f_1 \cdots f_n; \gamma).$$

4.4.3 Abstract Point of View on Wick Ordering

Definition 4.4.3.1 (*Gaussian Hilbert space*). A *Gaussian Hilbert space* $\mathcal{H}$ is a Hilbert space of random variables on some probability space $(\Omega, \sigma(\Omega), \mu)$ such that each $f \in \mathcal{H}$ is a Gaussian on $\mathbb{R}$, i.e., if $f \in \mathcal{H}$, then $f_*\mu$ is Gaussian on $\mathbb{R}$.

Example 4.4.3.1 If K denotes the completion of W^* in $L^2(W, \mu)$, then K is a Gaussian Hilbert space. We assume that $\sigma(\Omega)$ is generated by elements of $\mathcal{H}$.

Given a Gaussian Hilbert space $\mathcal{H} \subseteq L^2(\Omega, \sigma(\Omega), \mu)$, define the set

$$P_n(\mathcal{H}) := \{P(\xi_1, ..., \xi_n) \mid \xi_1, ..., \xi_n \in \mathcal{H}, P \text{ is a polynomial of degree } \leq n\},$$

and define

$$\mathcal{H}^{:n:} := \overline{P_n(\mathcal{H})} \cap (\overline{P_{n-1}(\mathcal{H})})^\perp.$$

We can then observe[7] the following points:

(1) We have

$$\overline{P_n(\mathcal{H})} = \bigoplus_{k=0}^{n} \mathcal{H}^{:k:},$$

(2) We have

$$\bigoplus_{n=0}^{\infty} \mathcal{H}^{:n:} = \overline{\bigcup_{n=0}^{\infty} P_n(\mathcal{H})}.$$

Theorem 4.4.3.1 *Let everything be as before. Then we get*

$$L^2(\Omega, \sigma(\Omega), \mu) = \bigoplus_{n=0}^{\infty} \mathcal{H}^{:n:}.$$

Proof See e.g., Anson (1987) for a proof. □

Remark 4.4.3.1 This is just a way to say that "polynomial" random variables are dense in $L^2(\Omega, \sigma(\Omega), \mu)$.

Theorem 4.4.3.2 *Let everything be as before and let $\xi_1, ..., \xi_n \in \mathcal{H}$. Then we get that*

$$: \xi_1 \cdots \xi_n := \pi_n(\xi_1 \cdots \xi_n),$$

where $\pi_n : L^2(\Omega, \sigma(\Omega), \mu) \to \mathcal{H}^{:n:}$ is the projection map.

Hence, it means that $: \xi_1, ..., \xi_n :$ is nothing but an orthogonal projection of $\xi_1 \cdots \xi_n$ onto $\mathcal{H}^{:n:}$. The idea here is that $\xi_1 \cdots \xi_n$ is not orthogonal to lower degree polynomials

---

[7] We will define the *completed* direct sum at some later point.

whereas : $\xi_1 \cdots \xi_n$: is orthogonal to lower degree polynomials. Hence, Wick ordering can be thought of as taking a polynomial and changing it into a new polynomial in such a way that the result is orthogonal to lower degree polynomials.

4.5 Bosonic Fock Spaces

Let $\mathcal{H}_1, \mathcal{H}_2$ be two separable Hilbert spaces. Then we can look at the tensor product $\mathcal{H}_1 \otimes \mathcal{H}_2$, where we have the inner product defined as

$$\langle h_1 \otimes h_2, h_1' \otimes h_2' \rangle_{\mathcal{H}_1 \otimes \mathcal{H}_2} := \langle h_1, h_1' \rangle_{\mathcal{H}_1} \langle h_2, h_2' \rangle_{\mathcal{H}_2}.$$

Moreover, we denote by $\mathcal{H}_1 \hat{\otimes} \mathcal{H}_2$ the completion of $\mathcal{H}_1 \otimes \mathcal{H}_2$ with respect to $\langle \cdot, \cdot \rangle_{\mathcal{H}_1 \otimes \mathcal{H}_2}$. We call $\hat{\otimes}$ the *Hilbert–Schmidt tensor product*. The space $\mathcal{H}_1^* \hat{\otimes} \mathcal{H}_2$ is isomorphic to the space of Hilbert–Schmidt operators from $\mathcal{H}_1$ to $\mathcal{H}_2$. Let $\{\mathcal{H}_i\}_{i=0}^\infty$ be a family of Hilbert spaces. Then $\widehat{\bigotimes}_i \mathcal{H}_i$ is the completion of $\bigotimes_i \mathcal{H}_i$ with respect to the norm $\sum_i \|x_i\|_{\mathcal{H}_i}^2$, i.e., we have

$$\widehat{\bigotimes}_i \mathcal{H}_i = \left\{ (x_i)_{i \geq 0} \,\middle|\, \sum_{i=0}^\infty \|x_i\|_{\mathcal{H}_i}^2 < \infty \right\}.$$

From now on, we will drop the "hat" and assume that the tensor product is completed. Note that more generally we can also define the space $\mathcal{H}_1 \otimes \cdots \otimes \mathcal{H}_n$ and so on. Let now $\mathcal{H}$ be a real and separable Hilbert space. Then, we can define its tensor power $T^n(\mathcal{H}) := \mathcal{H}^{\otimes n}$. Moreover, define the map $P : \mathcal{H}^{\otimes n} \to \mathcal{H}^{\otimes n}$ by $P(h_1 \otimes \cdots \otimes h_n) = \frac{1}{n!} \sum_{\sigma \in S_n} h_{\sigma(1)} \otimes \cdots \otimes h_{\sigma(n)}$. Then we can observe that P is a projection and we define $\mathrm{Sym}^n(\mathcal{H}) := P(\mathcal{H}^{\otimes n})$ to be the symmetric power of $\mathcal{H}$. Now we can see that the symmetric group S_n acts on the tensor power $T^n(\mathcal{H})$ and thus the symmetric power $\mathrm{Sym}^n(\mathcal{H})$ is the invariant subspace of $T^n(\mathcal{H})$ under this action. Given $h_1, ..., h_n \in \mathcal{H}$, we define the symmetric tensor product $\otimes_s$ by

$$h_1 \otimes_s \cdots \otimes_s h_n := \frac{1}{\sqrt{n!}} \sum_{\sigma \in S_n} h_{\sigma(1)} \otimes \cdots \otimes h_{\sigma(n)},$$

i.e., we have $h_1 \otimes_s \cdots \otimes_s h_n = \sqrt{n!} P(h_1 \otimes \cdots \otimes h_n)$.

Exercise 4.5.0.1 Show that, for $h_1, ..., h_n, h_1', ..., h_n' \in \mathcal{H}$, we have

$$\langle h_1 \otimes_s \cdots \otimes_s h_n, h_1' \otimes_s \cdots \otimes_s h_n' \rangle_{\mathcal{H}^{\otimes n}} = \sum_{\sigma \in S_n} \prod_{i=1}^n \langle h_i, h_{\sigma(i)}' \rangle_{\mathcal{H}}.$$

In particular, we have a relation to the norm on $\mathcal{H}$ given by $\|h^{\otimes_s n}\|^2_{\mathcal{H}^{\otimes_s n}} = n!\|h\|^{2n}_{\mathcal{H}}$, or, when taking the square root, $\|h^{\otimes_s n}\|_{\mathcal{H}^{\otimes_s n}} = \sqrt{n!}\|h\|_{\mathcal{H}}$.

Let us give an alternative definition for the symmetric power $\mathrm{Sym}^n(\mathcal{H})$. For this, recall that closed subspaces are generated by elements of the form $h_1 \otimes_s \cdots \otimes_s h_n$.

Remark 4.5.0.1 From now on we will not indicate the inner products.

Definition 4.5.0.1 (*Bosonic Fock space*). The space $\widetilde{\mathrm{Sym}}^{\bullet}(\mathcal{H}) := \bigoplus_{n=0}^{\infty} \mathrm{Sym}^n(\mathcal{H})$ is called the *bosonic Fock space* of $\mathcal{H}$.

Remark 4.5.0.2 We sometimes also write $\Gamma(\mathcal{H})$ or $\mathrm{Exp}(\mathcal{H})$ for the bosonic Fock space.

Remark 4.5.0.3 Similarly, one can define the *fermionic Fock space* of $\mathcal{H}$ by using elements of the form

$$u_1 \wedge \cdots \wedge u_n := \frac{1}{\sqrt{n!}} \sum_{\sigma \in S_n} \mathrm{sgn}(\sigma) u_{\sigma(1)} \otimes \cdots \otimes u_{\sigma(n)},$$

where $\wedge$ denotes the alternating tensor product.

One can check that there is indeed the *functorial property* $\Gamma(\mathcal{H}_1 \otimes \mathcal{H}_2) = \Gamma(\mathcal{H}_1) \otimes \Gamma(\mathcal{H}_2)$. Now one can ask whether there is an actual *functor* $\mathcal{H} \mapsto \Gamma(\mathcal{H})$. Thus, given a bounded operator $A \colon \mathcal{H}_1 \to \mathcal{H}_2$, we need to know whether $\Gamma(A)$ is bounded. As a matter of fact, this is not the case. On the other hand, if $\|A\| \leq 1$ then $\|\Gamma(A)\| \leq 1$. This leads to the functor

$$\Gamma \colon \mathbf{Hilb}_{\mathbf{B}}^{\leq 1} \longrightarrow \mathbf{Hilb},$$

where $\mathbf{Hilb}_{\mathbf{B}}^{\leq 1}$ is the category of Hilbert spaces together with bounded linear operators of norm ≤ 1 as morphisms and $\mathbf{Hilb}$ the category of Hilbert spaces with linear operators as morphisms.

Remark 4.5.0.4 There is actually no restriction required on A for the fermionic case.

Given $h \in \mathcal{H}$, we can define an *exponential map* on $\mathcal{H}$ by

$$\mathrm{Exp}(h) := \sum_{n=0}^{\infty} \frac{h^{\otimes n}}{n!} \in \Gamma(\mathcal{H}).$$

Then, for $h_1, h_2 \in \mathcal{H}$, we observe that

$$\langle \mathrm{Exp}(h_1), \mathrm{Exp}(h_2) \rangle = \sum_{n=0}^{\infty} \frac{1}{(n!)^2} \langle h_1^{\otimes_s n}, h_2^{\otimes_s n} \rangle = \sum_{n=0}^{\infty} \frac{1}{n!} \langle h_1, h_2 \rangle^n = \exp(\langle h_1, h_2 \rangle).$$

Exercise 4.5.0.2 Show that $\mathrm{Exp}\colon \mathcal{H} \to \Gamma(\mathcal{H})$ is continuous.[8] Moreover, show that the map Exp is injective.

Lemma 4.5.0.1 *The elements of the set* $\{\mathrm{Exp}(h) \mid h \in \mathcal{H}\} \subseteq \Gamma(\mathcal{H})$ *are linearly independent in* $\Gamma(\mathcal{H})$.

Proof Let $h_1, ..., h_n \in \mathcal{H}$. We want to show that if

$$\sum_{i=1}^{n} \lambda_i \mathrm{Exp}(h_i) = 0$$

for some $\lambda_i \in \mathbb{C}$, then $\lambda_i = 0$ for all $i = 1, ..., n$. For this, choose $h \in \mathcal{H}$ such that

$$\langle h, h_i \rangle_{\mathcal{H}} \neq \langle h, h_j \rangle_{\mathcal{H}}, \quad \forall i \neq j.$$

Then we get that $\sum\limits_{i=1}^{n} \lambda_i \mathrm{Exp}(h_i) = 0$ implies $\sum\limits_{i=1}^{n} \lambda_i e^{\langle h_i, h \rangle_{\mathcal{H}}} = 0$ for all $h \in \mathcal{H}$. Thus, we get that $\sum\limits_{i=1}^{n} \lambda_i e^{z \langle h_i, h \rangle_{\mathcal{H}}} = 0$ for all $z \in \mathbb{C}$ if we choose h as above. Hence, we get that $\lambda_i = 0$ for all $i = 1, ..., n$. $\qquad\square$

Exercise 4.5.0.3 Show that $\Gamma(\mathcal{H}) = \mathrm{span}\{\mathrm{Exp}(h) \mid h \in \mathcal{H}\}$.

Let us recall that

(1) $L^2(W, \mu) = \bigoplus\limits_{n=0}^{\infty} K^{:n:}$,

(2) there is a canonical isomorphism of Hilbert spaces $T\colon \mathcal{H}(\mu) \xrightarrow{\sim} K$.

Thus, we can observe that there is a canonical isomorphism of Hilbert spaces

$$\mathrm{Sym}^n(K) \longrightarrow K^{:n:},$$
$$\xi_1 \otimes_s \cdots \otimes_s \xi_n \longmapsto\, :\xi_1 \cdots \xi_n:,$$

which leads to a map

$$\Gamma(K) \longrightarrow \bigoplus_{n=0}^{\infty} K^{:n:}$$
$$\mathrm{Exp}(\xi) \longmapsto \sum_{n=0}^{\infty} \frac{:\xi^{\otimes n}:}{n!} =: \mathrm{Exp}(\xi) := e^{\xi} e^{-\frac{1}{2}\|\xi\|^2},$$

---

[8] However, it should be clear that it is *not* linear.

that comes from the *Segal–Ito isomorphism*

$$\Gamma(\mathcal{H}) \longrightarrow \Gamma(K) \cong L^2(W, \mu),$$
$$\mathrm{Exp}(h) \longmapsto e^{\langle h, \cdot \rangle} e^{-\frac{1}{2}\|h\|^2}.$$

Chapter 5
Construction of Quantum Field Theories

We want to understand the mathematical construction of the quantum theory of *fields*. There are several ways in which one can define such a theory and different approaches for its description. In particular, we are interested in the functional integral approach as we have seen for the special case of quantum mechanics, which is a 1-dimensional field theory.[1] Moreover, similarly to the case of quantum mechanics, we would like to formulate *axioms* in order to fully describe a quantum field theory.

5.1 Free Scalar Field Theory

A *(massive) classical scalar field theory* on the Euclidean space $\mathbb{R}^n$ consists of the following data: A *space of fields* $\mathcal{F} = C^\infty(\mathbb{R}^n)$ and an *action functional*, which is a map $S\colon \mathcal{F} \to \mathbb{R}$ that is *local*, i.e., it only depends on fields and higher derivatives of fields. In the *free theory* case we are interested in the action functional S which is of the form

$$S[\phi] = \int_{\mathbb{R}^n} \phi(\Delta + m^2)\phi \, \mathrm{d}x,$$

where $\mathrm{d}x$ denotes the Lebesgue measure on $\mathbb{R}^n$, the operator Δ denotes the Laplacian on $\mathbb{R}^n$ and m denotes the *mass* of the particle in the scalar field. As we have seen, for the quantum theory, we are interested in defining a measure on $\mathcal{F}$ of the form

$$e^{-\frac{1}{2}S[\phi]}\mathscr{D}[\phi]. \tag{5.1.1}$$

We will see that it is indeed possible to define a measure of the form (5.1.1), but it lives on a much larger space than $\mathcal{F}$.

---

[1] The one dimension considered there is *time*.

© The Author(s), under exclusive license to Springer Nature Singapore Pte Ltd. 2023
N. Moshayedi, *Quantum Field Theory and Functional Integrals*,
SpringerBriefs in Physics, https://doi.org/10.1007/978-981-99-3530-7_5

First, we will discuss Gaussian measures on locally convex spaces and as a consequence we will see how we can define a measure of the form (5.1.1).

5.1.1 Locally Convex Spaces

Definition 5.1.1.1 (Separating points). Let V be a vector space. A family $\{\rho_\alpha\}_{\alpha \in A}$ of semi-norms on V is said to *separate points* if $\rho_\alpha(x) = 0$ for all $\alpha \in A$ implies $x = 0$.

Definition 5.1.1.2 (Natural topology). Given a family of semi-norms $\{\rho_\alpha\}_{\alpha \in A}$ on V, there exists a smallest topology for which each ρ_α is continuous and the operation given by addition is continuous. This topology, which is denoted by $\mathcal{O}(\{\rho_\alpha\})$, is called the *natural topology* on V.

Definition 5.1.1.3 (Locally convex space). A *locally convex space* is a vector space V together with a family $\{\rho_\alpha\}_{\alpha \in A}$ of semi-norms that separates points.

Exercise 5.1.1.1 Show that the natural topology on a locally convex space is Hausdorff.

Let $\varepsilon > 0$ and $\alpha_1, ..., \alpha_n \in A$. Define then the set

$$N(\alpha_1, ..., \alpha_n; \varepsilon) := \{v \in V \mid \rho_{\alpha_i}(v) < \varepsilon, \ \forall i = 1, ..., n\}.$$

One can check that

(1) $N(\alpha_1, ..., \alpha_n; \varepsilon) = \bigcap_{i=1}^{n} N(\alpha_i; \varepsilon).$

(2) $N(\alpha_1, ..., \alpha_n; \varepsilon)$ is convex.

Exercise 5.1.1.2 Check that the elements of

$$\{N(\alpha_1, ..., \alpha_n; \varepsilon) \mid \alpha_1, ..., \alpha_n \in A, n \in \mathbb{N}, \varepsilon > 0\}$$

form a neighborhood basis at $0 \in V$.

From (2) it follows that a locally convex space V has a neighborhood basis at $0 \in V$, where each open set in this basis is convex. This justifies the name *locally convex space*. One can define the notion of a Cauchy sequence and the notion of convergence in a locally convex space. Let V be a locally convex space. The following are equivalent:

(1) V is metrizable.

(2) $0 \in V$ has a countable neighborhood basis.

(3) The natural topology on V is generated by some countable family of semi-norms.

Definition 5.1.1.4 (Fréchet space). A complete metrizable locally convex space is called a *Fréchet space*.

Example 5.1.1.1 (Schwartz space). One can easily check that $\mathcal{S}(\mathbb{R}^n)$, defined as in Sect. 3.3.2, is a locally convex space. In general $\mathcal{S}(\mathbb{R}^n)$ is a Fréchet space.

5.1.2 Dual of a Locally Convex Space

Let V be a locally convex space and V^* be the set of continuous linear functionals on V, i.e., $\ell \in V^*$ if and only if $\ell \colon V \to \mathbb{R}$ is linear and continuous. Given $x \in V$, define a map $\rho_x \colon V^* \to \mathbb{R}$ by $\rho_x(\ell) = |\ell(x)|$. One can easily check that ρ_x is a semi-norm. In fact $\{\rho_x \mid x \in V\}$ is a family of semi-norms on V^* that separates points. Hence, $(V^*, \{\rho_x \mid x \in V\})$ is a locally convex space. The natural topology on V^* induced by $\{\rho_x \mid x \in V\}$ is called the *weak*-topology* on V^*. A sequence $\{\ell_n\}$ in V^* converges to $\ell \in V^*$ in the weak*-topology if and only if $\rho_x(\ell_n) \xrightarrow{n \to \infty} \rho_x(\ell)$ for all $x \in V$, i.e., $\ell_n(x) \xrightarrow{n \to \infty} \ell(x)$ for all $x \in V$. The weak*-topology on V^* is denoted by $\mathcal{O}(V^*, V)$.

Remark 5.1.2.1 Note that the space of linear functionals on $(V^*, \mathcal{O}(V^*, V))$ is exactly V.

5.1.3 Gaussian Measures on the Dual of Fréchet Spaces

Theorem 5.1.3.1 *Let V be a Fréchet space. Then there is a bijection between the following sets:*

$$\Big\{ \textit{Continuous, positive-definite, symmetric bilinear forms on } V \Big\}$$
$$\longleftrightarrow \Big\{ \textit{Centered Gaussian measures on } (V^*, \mathcal{O}(V^*, V)) \Big\}.$$

Proof See Bogachev (1998), Guichardet (1972) for a proof. $\qquad\square$

Let C be a continuous, positive-definite, symmetric bilinear form on V. The construction of the associated Gaussian measure on V^* works as follows. Let $F \subseteq V$ be a finite-dimensional subspace of V. Let C_F be the restriction of C on F. Then C_F is a symmetric bilinear form on F, which is positive-definite. Hence, we get that C_F defines a Gaussian measure μ_{C_F} on[2] F^* of the form

$$d\mu_{C_F}(x) = \left(\det\left(\frac{C_F}{2\pi} \right) \right)^{-\frac{1}{2}} e^{-\frac{1}{2}C_F(x,x)},$$

---

[2] This means that F^* can be identified with F.

where C_F is identified with a positive-definite matrix. In fact, one can think of μ_{C_F} to be a measure on the F-cylinder subsets of V^*. One can check that if $E \subseteq F$, then μ_{C_F} agrees with μ_{C_E} when restricted to the E-cylinder subsets of V^*. Now we can proceed as in the construction of the Wiener measure and show that there is a Gaussian measure μ_C on the sigma algebra of V^* generated by cylinder sets. This gives the construction of μ_C.

Corollary 5.1.3.1 *Let C be the bilinear form on $\mathcal{S}(\mathbb{R}^n)$ defined by*

$$C(f, g) = \int_{\mathbb{R}^n} f(\Delta + m^2)^{-1} g \, dx.$$

Then there is a Gaussian measure μ on $\mathcal{S}(\mathbb{R}^n)$ whose covariance is C.

In this case the reproducing kernel space $K(\mu)$ of μ is $H^{-1}(\mathbb{R}^n)$, where $H^{-1}(\mathbb{R}^n)$ is the completion of $\mathcal{S}(\mathbb{R}^n)$ with respect to C. Hence, we are able to define a measure of the form $e^{-S[\phi]} \mathscr{D}[\phi]$, where

$$S[\phi] = \frac{1}{2} \int_{\mathbb{R}^n} \phi(\Delta + m^2) \phi \, dx.$$

In other words, we can construct the Gaussian measure associated with the free theory. In this case, the Cameron–Martin space of μ is $H^1(\mathbb{R}^n)$, which is the completion of $\mathcal{S}(\mathbb{R}^n)$ with respect to the map

$$(f, g) \mapsto \int_{\mathbb{R}^n} f(\Delta + m^2) g \, dx.$$

5.1.4   The Operator $(\Delta + m^2)^{-1}$

In order to understand the reproducing kernel, we have to understand the operator $(\Delta + m^2)^{-1}$, which we will regard as an operator on $L^2(\mathbb{R}^n)$. It is well-known that $(\Delta + m^2)^{-1}$ is a positive and integral operator. Let $C(x, y)$ be the integral kernel of $(\Delta + m^2)^{-1}$. Then we get that

$$C(f, g) = \iint_{\mathbb{R}^n \times \mathbb{R}^n} f(x) C(x, y) g(y) \, dx \, dy.$$

In fact, one can show that $C(x, y)$ is the unique solution of the partial differential equation

$$\Delta_y C(x, y) = \delta_x(y). \tag{5.1.2}$$

Using Fourier transform, we can give an explicit representation of $C(x, y)$. Formally, we have the following chain of implications:

$$(\Delta + m^2)^{-1}f = g \Rightarrow (\Delta + m^2)g = f \Rightarrow (\xi^2 + m^2)\mathcal{F}(g) = \mathcal{F}(f) \Rightarrow g = \mathcal{F}^{-1}\left(\frac{1}{\xi^2 + m^2}\mathcal{F}(f)\right).$$

These implications hold since

$$g(x) = \frac{1}{(2\pi)^n}\iint_{\mathbb{R}^n \times \mathbb{R}^n}\frac{e^{i\xi(x-y)}}{\xi^2 + m^2}f(y)\,dy\,d\xi,$$

and thus we get

$$C(x, y) = \frac{1}{(2\pi)^n}\int_{\mathbb{R}^n}\frac{e^{i\xi(x-y)}}{\xi^2 + m^2}\,d\xi.$$

For $x \neq y$ one can actually show that

$$C(x, y) = (2\pi)^{-\frac{1}{2}}\left(\frac{m}{\|x - y\|}\right)^{\frac{n-2}{2}}K_{\frac{n-2}{2}}(m\|x - y\|),$$

where K_ν is a modified *Bessel function*. Next, we will study $C(x, y)$ in more detail. In particular, we want to study the behaviour of $C(x, y)$ when $\|x - y\|$ is large and when $\|x - y\|$ is small.

Remark 5.1.4.1 In particular, for $n = 1$ we have $C(x, y) = \frac{1}{m}e^{-m|x-y|}$ and for $n = 3$ we have $C(x, y) = \frac{1}{4\pi\|x-y\|}e^{-m\|x-y\|}$.

Proposition 5.1.4.1 (Properties of $C(x, y)$). *Let everything be as before. Then the following hold:*

(1) *For every $m|x - y|$ bounded away from zero, there exists some $M \geq 0$ such that we have*
$$C(x, y) \leq Mm^{\frac{n-3}{2}}|x - y|^{\frac{n-1}{2}}e^{-m|x-y|}.$$

(2) *For $n \geq 3$ and $m|x - y|$ in a neighborhood of zero we get*
$$C(x, y) \sim |x - y|^{-n+2}.$$

(3) *For $n = 2$ and $m|x - y|$ in a neighborhood of zero we get*
$$C(x, y) \sim -\log(m|x - y|).$$

Proof Let us first recall that

$$C(x, y) = \frac{1}{2\pi}\int_{\mathbb{R}}\frac{e^{i\xi(x-y)}}{\xi^2 + m^2}\,d\xi. \tag{5.1.3}$$

Thus we can also show the following exercise.

Exercise 5.1.4.1 Show that in general we have

$$C(x, y) = \frac{1}{(2\pi)^n} \int_{\mathbb{R}^n} \frac{e^{i\xi\|x-y\|}}{\xi^2 + m^2} d\xi.$$

Hint: Choose an orthonormal basis $\{e_1, ..., e_n\}$ of $\mathbb{R}^n$ such that $e_1 = \frac{x-y}{\|x-y\|}$ and do a change of variables.

Now, using Exercise 5.1.4.1 and the residue theorem, we get that

$$C(x, y) = \frac{1}{(2\pi)^n} \int_{\mathbb{R}^n} \frac{e^{it\xi_1}}{\xi_1^2 + \left(\sqrt{m^2 + \xi_2^2 + \cdots + \xi_n^2}\right)^2} d\xi_1 d\xi_2 \cdots d\xi_n$$

$$= \frac{1}{(2\pi)^n} \int_{\mathbb{R}^{n-1}} \frac{\pi e^{-t\sqrt{m^2+\xi_2^2+\cdots+\xi_n^2}}}{\sqrt{m^2 + \xi_2^2 + \cdots + \xi_n^2}} d\xi_2 \cdots d\xi_n.$$

Without loss of generality, assume that $m = 1$. Recall that for a function $f: \mathbb{R}^n \to \mathbb{R}$ with $f(x) = g(|x|)$ we have

$$\int_{\mathbb{R}^n} f(x)dx = v(S^{n-1}) \int_0^\infty r^{n-1} g(r)dr, \tag{5.1.4}$$

where $v(S^{n-1})$ denotes the volume of S^{n-1}. Using (5.1.4), we can write

$$C(x, y) = \frac{\pi A_{n-1}}{(2\pi)^{\frac{n}{2}}} \int_0^\infty \frac{r^{n-2} e^{-t\mu(r)}}{\mu(r)} dr, \tag{5.1.5}$$

where $\mu(r) := \sqrt{1 + r^2}$.

Exercise 5.1.4.2 Show that there is some $\varepsilon > 0$ such that

$$\mu(r) \geq \begin{cases} 1 + \varepsilon r^2, & \text{if } r \leq 1, \\ 1 + \varepsilon r, & \text{if } r \geq 1. \end{cases}$$

Next, we claim that
$$C(x, y) \leq k e^{-t}\left(t^{-\frac{n-1}{2}} + t^{-(n-1)}\right), \tag{5.1.6}$$

where $t = \|x - y\|$ and k is some constant. Note that then we get

$$\int_0^1 \frac{r^{n-2} e^{-t\mu(r)}}{\mu(r)} \leq \int_0^1 r^{n-2} e^{-t(1+\varepsilon r^2)} dr \leq k e^{-t} t^{-\frac{n-1}{2}},$$

and

$$\int_1^\infty \frac{r^{n-2}e^{-t\mu(r)}}{\mu(r)} \le \int_1^\infty r^{n-2}e^{-t(1+\varepsilon r)}dr \le ke^{-t}t^{-(n-1)}.$$

If $t \ge 1$, we get that

$$C(x, y) \le ke^{-t}t^{-\frac{n-1}{2}}.$$

For $0 < t \le 1$, we get that

$$\int_0^\infty r^{n-2}\frac{e^{-t\sqrt{1+r^2}}}{1+r^2}dr = t^{-(n-2)}\int_0^\infty s^{n-2}\frac{e^{-\sqrt{s^2+t^2}}}{\sqrt{s^2+t^2}}ds$$

$$\sim_{(t\to 0)} t^{-(n-2)}\int_0^\infty \frac{s^{n-2}e^{-s}}{s}ds$$

$$= t^{-(n-2)}\int_0^\infty s^{n-3}e^{-s}ds.$$

For $n = 2$, define $s := t\mu(r)$. Then we get that $\mu(r) = \frac{s}{t}$, $1 + r^2 = \frac{s^2}{t^2}$ and $r = \sqrt{\frac{s^2-t^2}{t^2}}$. Thus, we get

$$C(x, y) = \int_t^\infty \frac{e^{-s}}{\sqrt{s^2+t^2}}ds \sim \int_t^\infty \frac{1}{\sqrt{s^2+t^2}}ds \sim -\log(t).$$

$\square$

5.2 Construction of Self-Interacting Theories

To construct a theory with *polynomial interaction*, we want to rigorously define a measure of the form

$$e^{-S[\phi]}\mathscr{D}[\phi],$$

where

$$S[\phi] = \frac{1}{2}\int_{\mathbb{R}^n} \phi(\Delta + m^2)\phi dx + \int_{\mathbb{R}^n} P(\phi)dx = S_{\text{free}}[\phi] + S_{\text{int}}[\phi],$$

such that $P(y) = \sum_i a_i y^i$ is some polynomial function on $\mathbb{R}$. We were already able to define a measure μ of the form $e^{-S_{\text{free}}[\phi]}\mathscr{D}[\phi]$, but with the restriction that it is defined on $\mathcal{S}(\mathbb{R}^n)$. In fact, we have $\mu(\mathcal{S}(\mathbb{R}^n)) = 0$ because for such measures the Cameron–Martin space is $H^1(\mathbb{R}^n)$ and we have $\mathcal{S}(\mathbb{R}^n) \subseteq H^1(\mathbb{R}^n)$. Thus, it is not obvious that we have to regard $\phi \mapsto \int_{\mathbb{R}^n} \phi^n(x)dx$ as a measurable function on $\mathcal{S}(\mathbb{R}^n)$. Let us now define a measure of the form

$$e^{-S_{\text{free}}[\phi]}e^{-S_{\text{int}}[\phi]}\mathscr{D}[\phi].$$

First, we will "define" measurable functions of the form

$$\phi \longmapsto \int_{\mathbb{R}^n} \phi(x)^k \mathrm{d}x. \tag{5.2.1}$$

In particular, we want to bypass the difficulties in making sense of (5.2.1). Let us pretend that we can define (5.2.1). Formally, we then have

$$\left\| \int_{\mathbb{R}^n} \phi(x)^k \mathrm{d}x \right\|^2_{L^2(\mathcal{S}(\mathbb{R}^n),\mu)} = \int_{\phi \in \mathcal{S}(\mathbb{R}^n)} \left(\int_{\mathbb{R}^n} \phi(x)^k \mathrm{d}x \right)\left(\int_{\mathbb{R}^n} \phi(y)^k \mathrm{d}y \right) \mathrm{d}\mu(\phi)$$

$$= \iint_{\mathbb{R}^n \times \mathbb{R}^n} \left(\int_{\phi \in \mathcal{S}(\mathbb{R}^n)} \phi(x)^k \phi(y)^k \mathrm{d}\mu(\phi) \right) \mathrm{d}x\mathrm{d}y.$$

Now, formally, think of $\phi(x)$ as $\delta_x[\phi] = \langle \delta_x | \phi \rangle$. Then, using Wick's theorem (Theorem 4.3.2.1), we see that

$$\left\| \int_{\mathbb{R}^n} \phi(x)^k \mathrm{d}x \right\|^2_{L^2(\mathcal{S}(\mathbb{R}^n),\mu)} = \text{Linear combination of integrals of the form}$$

$$\iint_{\mathbb{R}^n \times \mathbb{R}^n} C(x,x)^\alpha C(y,y)^\beta C(x,y)^\gamma \mathrm{d}x\mathrm{d}y.$$

The existence of $\int_{\mathbb{R}^n} \phi(x)^k \mathrm{d}x$ depends on the properties of $C(x,y)$ and hence it is *dimension sensitive*. In fact, we cannot define (5.2.1) because $C(x,x)$ is not integrable. We want to try two different attempts to make sense of $\int_{\mathbb{R}^n} \phi(x)^k \mathrm{d}x$. In particular, we want to consider the following two points:

(1) (Approximation of delta function) Let $h \in \mathcal{S}(\mathbb{R}^n)$ such that $h \geq 0, h(0) > 0$ and $\int_{\mathbb{R}^n} h\mathrm{d}x = 1$. E.g., consider $h_\varepsilon(y) := \frac{1}{\varepsilon^n} h\left(\frac{y}{\varepsilon}\right)$ for some $\varepsilon > 0$. Then we get that $h_\varepsilon \xrightarrow{\varepsilon \to 0} \delta_0$. Let $\mathcal{S}'(\mathbb{R}^n)$ denote the space of *Schwartz distributions* on $\mathbb{R}^n$. Similarly[3] as before, we can construct $\delta_{\varepsilon,x} \in \mathcal{S}'(\mathbb{R}^n)$ such that $\delta_{\varepsilon,x} \xrightarrow{\varepsilon \to 0} \delta_x$. Thus, we get that the map $\phi \mapsto \delta_{\varepsilon,x}[\phi] = \langle \delta_{\varepsilon,x} | \phi \rangle$ is a polynomial function on $\mathcal{S}(\mathbb{R}^n)$. We denote the polynomial by $\phi(\delta_{\varepsilon,x})$. Actually, we already know how to compute an integral of the form

$$\int_{\phi \in \mathcal{S}(\mathbb{R}^n)} \phi(\delta_{\varepsilon,x})^k \phi(\delta_{\varepsilon,y})^m \mathrm{d}\mu(\phi).$$

It is equal to the sum of terms of the form $A_{\alpha\beta\gamma} C(\delta_{\varepsilon,x}, \delta_{\varepsilon,x})^\alpha C(\delta_{\varepsilon,y}, \delta_{\varepsilon,y})^\beta$ $C(\delta_{\varepsilon,x}, \delta_{\varepsilon,y})^\gamma$ for some constants $A_{\alpha\beta,\gamma}$. Formally, we have that the inner product

---

[3] With $k = \frac{1}{\varepsilon}$, we get that $h_k(x) = k^2 h(kx) \xrightarrow{k \to \infty} \delta_0$.

$$\left\langle \int_{\mathbb{R}^n} \phi(\delta_{\varepsilon,x})^k \mathrm{d}x, \int_{\mathbb{R}^n} \phi(\delta_{\varepsilon,y})^m \right\rangle_{L^2(\mathcal{S}(\mathbb{R}^n),\mu)}$$

is equal to a sum of expressions of the form

$$A_{\alpha\beta\gamma} \iint_{\mathbb{R}^n \times \mathbb{R}^n} C(\delta_{\varepsilon,x}, \delta_{\varepsilon,x})^\alpha C(\delta_{\varepsilon,y}, \delta_{\varepsilon,y})^\beta C(\delta_{\varepsilon,x}, \delta_{\varepsilon,y})^\gamma \mathrm{d}x\mathrm{d}y,$$

and one could try to take the limit $\varepsilon \to 0$. The conclusion here is that this attempt does not lead anywhere, since we still get a diagonal contribution.

(2) (Redefine observables) Let us try to get rid of diagonal contributions, i.e., terms proportional to $C(x, x)$. This is where the *Wick ordering* comes into the play. We can think of Wick ordering as a *renormalization* process.[4] Consider the map

$$\phi \longmapsto \int_{\substack{\Lambda \subset \mathbb{R}^n \\ \Lambda \text{ compact}}} :\phi(\delta_{\varepsilon,x})^k: \mathrm{d}x.$$

Thus, we get that

$$\iint_{\mathbb{R}^n \times \mathbb{R}^n} \left(\int_{\mathcal{S}(\mathbb{R}^n)} :\phi(\delta_{\varepsilon,x})^k :: \phi(\delta_{\varepsilon,y})^k: \mathrm{d}\mu(\phi) \right) \mathrm{d}x\mathrm{d}y = k! \iint_{\mathbb{R}^n \times \mathbb{R}^n} C(\delta_{\varepsilon,x}, \delta_{\varepsilon,y})^k \mathrm{d}x\mathrm{d}y.$$

Taking the limit $\varepsilon \to 0$, formally, it converges to

$$k! \iint_{\mathbb{R}^n \times \mathbb{R}^n} C(x, y)^k \mathrm{d}x\mathrm{d}y. \tag{5.2.2}$$

Let us make a short summary of what we have seen so far:

(i) Wick ordering allows us to get rid of diagonal contributions of $C(x, y)$.

(ii) If $\iint_{\mathbb{R}^n \times \mathbb{R}^n} C(x, y)^k \mathrm{d}x\mathrm{d}y < \infty$, there is hope that we can define "Wick ordered polynomial" functions of the form

$$\phi \longmapsto \int_{\mathbb{R}^n} :\phi(x)^k: \mathrm{d}x.$$

Recall that if $n \geq 3$, we get that $C(x, y) \sim \frac{1}{\|x-y\|^{n-2}}$ for $\|x - y\| \to 0$ and hence the integral of the form (5.2.2) will diverge in general. This means that if $n \geq 3$, Wick ordering renormalization may not get rid of all the infinities appearing in Feynman amplitudes. However, if $n = 2$, we get that $C(x, y) \sim -\log(\|x - y\|)$ as $\|x - y\| \to 0$ and in this case it is possible to define observables of the form $\phi \mapsto \int_{\substack{\Lambda \subset \mathbb{R}^2 \\ \Lambda \text{ compact}}} :P(\phi): \mathrm{d}x$. Thus, define $S_{\text{int},\Lambda}[P(\phi)] := \int_{\substack{\Lambda \subset \mathbb{R}^2 \\ \Lambda \text{ compact}}} :P(\phi): \mathrm{d}x$ as a measurable function on

---

[4] These are certain procedures to get rid of infinities such that the self-interacting theory becomes meaningful.

$S'(\mathbb{R}^2)$, where P is any polynomial. Consider then $P(x) = x^4$ and a compact subset $\Lambda \subset \mathbb{R}^2$. Then, we can define $S^{\varepsilon}_{\text{int},\Lambda}[P(\phi)] := \int_{\Lambda} : P(\phi, \delta_{\varepsilon,x}) : dx$, where $\delta_{\varepsilon,x}$ is a *smooth* approximation of δ_x. More precisely, $\delta_{\varepsilon,x}$ can be constructed as follows. Let $h \in C_0^{\infty}(\mathbb{R}^2)$ with $h \geq 0$, $h(0) > 0$, and $\int_{\mathbb{R}^2} h dx = 1$. Then, we define $\delta_{\varepsilon,x}(y) := \varepsilon^{-2} h\left(\frac{x-y}{\varepsilon}\right)$. In fact, we get that $\delta_{\varepsilon,x} \xrightarrow{\varepsilon \to 0} \delta_x$ in $S'(\mathbb{R}^2)$. If we take $\varepsilon = \frac{1}{k}$, we will get that $\delta_{\varepsilon,x} = \delta_{\frac{1}{k},x}$ and thus one can observe that the sequence $\left\{S^{1/k}_{\text{int},\Lambda}\right\}_{k>0}$ is Cauchy in $L^2(S'(\mathbb{R}^2), \mu)$. We then define $S_{\text{int},\Lambda}[P(\phi)] := \lim_{k \to \infty} S^{1/k}_{\text{int},\Lambda}[P(\phi)]$.

Remark 5.2.0.1 Recall that we have

$$\left\langle \int_{\Lambda} : \phi\left(\delta_{\frac{1}{k},x}\right) :^n, \int_{\Lambda} : \phi\left(\delta_{\frac{1}{k},y}\right) :^n \right\rangle_{L^2(S'(\mathbb{R}^2),\mu)} = n! \int_{\Lambda \times \Lambda} C\left(\delta_{\frac{1}{k},x}, \delta_{\frac{1}{k},y}\right)^n dx dy.$$

To see that $\left\{S^{1/k}_{\text{int},\Lambda}\right\}_{k>0}$ is Cauchy, we only have to understand how C behaves. In the Fourier picture it is actually easier to understand. We know that the Fourier transform of $\delta_{\frac{1}{k},x}$ is given by $\left(\frac{1}{\xi^2+m^2}\right)^2 \mathcal{F}(h)\left(\frac{\xi}{k}\right)$. Then, one can use the properties of $C_k(x, y) := C\left(\delta_{\frac{1}{k},x}, \delta_{\frac{1}{k},y}\right)$, which is a smooth approximation of the Green's function, to show that the sequence $\int_{\Lambda} : \phi\left(\delta_{\frac{1}{k},x}\right) :^n dx$ is Cauchy.

5.2.1 More Random Variables

Let us consider a function $f \in C_0^{\infty}(\underbrace{\mathbb{R}^2 \times \cdots \times \mathbb{R}^2}_{k}) = C^{\infty}(\mathbb{R}^{2k})$. Then, we get that

$$S_{\text{int},\Lambda}[\phi, f; k] := \int_{\underbrace{\mathbb{R}^2 \times \cdots \times \mathbb{R}^2}_{k}} : \phi(x_1) \cdots \phi(x_k) : f(x_1, ..., x_k) dx_1 \cdots dx_k$$

can be defined similarly as before. More generally, note that we can also take $f \in L^2(\mathbb{R}^2 \times \cdots \times \mathbb{R}^2)$. Moreover, for $f_i \in C_0^{\infty}(\mathbb{R}^{2k_i})$ with $i = 1, ..., n$, we can also define

$$A[\phi] := \prod_{i=1}^{n} S_{\text{int},\Lambda}[\phi, f_i; k_i].$$

We want to understand how we can compute the integral $\int_{S'(\mathbb{R}^2)} A[\phi] d\mu(\phi)$. For that, we recall the following statement: For any measure space (W, μ) and any linear map $f \in W^*$ we have

$$: f^n := \sum_{k=0}^{\lfloor \frac{n}{2} \rfloor} \frac{n!}{k!(n-2k)!} f^{n-2k} \left(-\frac{1}{2} q_\mu(f, f) \right)^k.$$

We would like to know if an expression of the form

$$: f_1 \cdots f_k :: g_{k+1} \cdots g_n :$$

can be written as a linear combination of Wick ordered polynomials. The answer is *yes*, and the advantage is that it allows us to simplify the computation of an integral consisting of products of Wick ordered polynomials.

Example 5.2.1.1 We can write

$$: f_1 \cdots f_n := f_1 : f_2 \cdots f_n : - \sum_{j=2}^{n} q_n(f_1, f_j) : f_2 \cdots \hat{f}_j \cdots f_n :,$$

which is similar to integration by parts. Here the symbol ˆ means that the element is omitted.

5.2.2 Generalized Feynman Diagrams

Let $I = I_1 \sqcup I_2 \sqcup \cdots \sqcup I_n$, be the disjoint union of finite sets I_i for $i \in \{1, ..., n\}$.

Definition 5.2.2.1 (Generalized Feynman diagram). A *generalized Feynman diagram* is a pair (I, E), where

$$E \subseteq \{(a, b) \mid a \text{ and } b \text{ do not belong to some } I_i \text{ for all } i \in \{1, ..., n\}\}$$

denotes the set of *edges*.

We denote by A_E the remaining vertices. Let $\gamma := (V, E)$ be a generalized Feynman diagram associated with $: f_1 \cdots f_k :: g_{k+1} \cdots g_n :$. Then, we get that the set of *vertices* of γ is given by

$$V(F) = \prod_{e \in E} q_\mu(f_{\ell(e)}, g_{r(e)}) : \prod_{v \in A_E} \alpha_v :,$$

where α_v is either f_v or g_v and $\ell(e)$ is the left end point and $r(e)$ the right end point of the edge E.

Corollary 5.2.2.1 *Let everything be as before. Then we get that*

$$\int : f_1 \cdots f_k :: g_{k+1} \cdots g_n : d\mu = \sum_{\substack{\gamma \text{ complete} \\ \text{Feynman diagram}}} F(: f_1 \cdots f_k :: g_{k+1} \cdots g_n :; \gamma).$$

Consider now again the integral $\int_{\mathcal{S}'(\mathbb{R}^2)} A[\phi] d\mu(\phi)$. It can be computed using generalized Feynman diagrams. Our goal is to show that $\mathrm{e}^{-S_{\mathrm{int},\Lambda}} \in L^1(\mathcal{S}'(\mathbb{R}^n))$. For this, let us define $S_{\mathrm{int},\Lambda}[\phi; 2k] := \int_{\Lambda} : \phi^{2k}(x) : \mathrm{d}x$. Now we want to check that $\mathrm{e}^{-S_{\mathrm{int},\Lambda}[\phi; 2k]} \in L^1(\mathcal{S}'(\mathbb{R}^n))$. If we consider e.g., $: x^4 := x^4 - 6x^2 + 3$, we can see that $\mathrm{e}^{-:x^4:}$ can indeed behave badly. However, we can show that the following lemma holds.

Lemma 5.2.2.1 *Let everything be as before. Then we get that*

$$S_{\mathrm{int},\Lambda}[\phi; 2k] \geq -b(\log k)^n,$$

as $k \to \infty$ for some $b > 0$.

Remark 5.2.2.1 This shows that $S_{\mathrm{int},\Lambda}[\phi; 2k]$ does not generalize to a polynomial, which is not bounded from below.

Proof Let $Q(y) := \sum\limits_{k=0}^{2n} a_k y^k$ with $a_{2n} > 0$. Then, we get that

$$\inf_{y \in \mathbb{R}} Q(y) \geq -b,$$

for some $0 \leq b < \infty$, and

$$: \phi\left(\delta_{\frac{1}{k},x}\right)^{2n} := \sum_{k=0}^{2n} \frac{(2n)!}{k!} \phi\left(\delta_{\frac{1}{k},x}\right)^{2n-2k} \left(-\frac{1}{2} C\left(\delta_{\frac{1}{k},x}, \delta_{\frac{1}{k},x}\right)\right)^k$$

$$= C_k(x,x)^n \sum_{k=0}^{2n} \frac{(2n)!}{k!(2n-2k)!} \frac{(-1)^k}{2^k} \left(\frac{\phi\left(\delta_{\frac{1}{k},x}\right)}{\sqrt{C_k(x,x)}}\right)^{2n-2k}.$$

Thus, we get that $: \phi\left(\delta_{\frac{1}{k},x}\right) :^{2n} \geq -b \int_{\Lambda} C_k(x,x)^n \mathrm{d}x$ for some $b > 0$ and hence $S_{\mathrm{int},\Lambda;k}[\phi] \geq -b \int_{\Lambda} C_k(x,x)^n \mathrm{d}x \geq -b(\log k)^n$ as $k \to \infty$. $\square$

Corollary 5.2.2.2 *Let everything be as before. Then we get that $\mathrm{e}^{-S_{\mathrm{int},\Lambda;k}} \in L^p(\mathcal{S}'(\mathbb{R}^n))$ for all $p \geq 0$.*

Consider now the integral $S_{\mathrm{int}}[f, P(\phi)] := \int_{\mathbb{R}^2} f(x) : P(\phi(x)) : \mathrm{d}x$, where $P(x) = \sum\limits_n a_n x^n$ is a polynomial function on $\mathcal{S}'(\mathbb{R}^2)$ and $f \in L^2(\mathbb{R}^2)$. We have already shown that $S_{\mathrm{int}}[f, P(\phi)] \in L^2(\mathcal{S}'(\mathbb{R}^2), \mu)$. Let $S_{\mathrm{int},\Lambda}^{1/k}[f, P(\phi)] := \int_{\Lambda} f(x) P\left(\phi, \delta_{\frac{1}{k},x}\right) \mathrm{d}x$, where $\delta_{\frac{1}{k},x}$ is a *smooth* approximation of δ_x. We have also seen that if $S_{\mathrm{int},\Lambda}^{1/k}[f, P(\phi)]$ is Cauchy, then $S_{\mathrm{int},\Lambda}^{1/k}[f, P(\phi)] = \lim\limits_{k \to \infty} S_{\mathrm{int},\Lambda}^{1/k}[f, P(\phi)]$. In fact, we get that

$$\left\| S_{\mathrm{int},\Lambda}^{1/k}[f, P(\phi)] - S_{\mathrm{int},\Lambda}[f, P(\phi)] \right\|_{L^2(\mathcal{S}'(\mathbb{R}^2),\mu)} \leq Cf^{-\delta},$$

for some $\delta > 0$ as $k \to \infty$, where $C > 0$ is some constant. Moreover, we get that $e^{-S_{\text{int},\Lambda}[P(\phi);2k]} \in L^1(\mathcal{S}'(\mathbb{R}^n))$, where $S_{\text{int},\Lambda}[P(\phi); 2k] := \int_\Lambda : P(\phi(x)) : \mathrm{d}x$ with $P(x) = x^{2k}$. The idea is that $S_{\text{int},\Lambda}[P(\phi); 2k] \geq -C(\log k)^n$ for some constant $C > 0$. Moreover, we can observe that $S_{\text{int},\Lambda}[P(\phi); 2k] \geq 1 - \tilde{C}(\log k)^n$ for some constant $\tilde{C} > 0$ for large k, e.g., we can define $\tilde{C} := \frac{2}{3}C$. Recall that the goal was to show that $e^{-S_{\text{int},\Lambda}[P(\phi);2k]} \in L^1(\mathcal{S}'(\mathbb{R}^2), \mu)$. The strategy here is to study the sets on which $S_{\text{int},\Lambda}[P(\phi); 2k]$ behaves badly, and then show that these sets have measure zero. Define a "bad set" to be given by

$$X(k) := \{\phi \in \mathcal{S}'(\mathbb{R}^2) \mid S_{\text{int},\Lambda}[P(\phi); 2k] \leq \tilde{C}(\log k)^n\}.$$

Lemma 5.2.2.2 *Let everything be as before. Then we get that*

$$X(k) \subseteq \{\phi \in \mathcal{S}'(\mathbb{R}^2) \mid \left\| S_{\text{int},\Lambda}[P(\phi)] - S_{\text{int},\Lambda}[P(\phi); 2k] \right\| \geq 1\}.$$

Proof Let $\phi \in X(k)$. Then we get that

$$S_{\text{int},\Lambda}[P(\phi)] - S_{\text{int},\Lambda}[P(\phi); 2k] \leq S_{\text{int},\Lambda}[P(\phi)] - (1 - \tilde{C}(\log k)^n)$$
$$= \underbrace{S_{\text{int},\Lambda}[P(\phi)] + \tilde{C}(\log k)^n}_{\leq 1} - 1.$$

$\square$

Proposition 5.2.2.1 *There is a constant $B_\Lambda > 0$ and $\delta > 0$ such that*

$$\mu(X(k)) \leq B_\Lambda k^{-\delta} \quad \text{as } k \to \infty.$$

Proof Note that we have

$$\mu(X(k)) = \int_{X(k)} \mathrm{d}\mu \leq \int_{X(k)} \| S_{\text{int},\Lambda}[P(\phi)] - S_{\text{int},\Lambda}[P(\phi); 2k] \|^2 \mathrm{d}\mu$$
$$\leq \int_{\mathcal{S}'(\mathbb{R}^2)} \| S_{\text{int},\Lambda}[P(\phi)] - S_{\text{int},\Lambda}[P(\phi); 2k] \|^2 \mathrm{d}\mu$$
$$\leq B_\Lambda k^{-\delta}$$

as $k \to \infty$. $\square$

Remark 5.2.2.2 One can actually show that $\mu(X(k)) \leq C\text{Exp}(-k^\alpha)$ for some $\alpha > 0$ as $k \to \infty$.

Let now $(\Omega, \sigma(\Omega), \mu)$ be a probability space and consider a measurable function $f: \Omega \to \mathbb{R}$. Moreover, define the set

$$\mu_f(x) := \mu(\{\omega \in \Omega \mid f(\omega) \geq x\}).$$

Let F be an increasing positive function on $\mathbb{R}$ such that $\lim\limits_{x\to\infty} F(x) = \infty$. Then, we get that

$$\int_{\Omega} F(f(\omega))d\mu(\omega) = \int_{\mathbb{R}} F(x)\mu_f(x)dx.$$

Theorem 5.2.2.1 *Let f be a measurable function on Ω such that*

$$\mu(\{\omega \in \Omega \mid -f(\omega) \geq C(\log k)^n\}) \leq Ce^{-k_0^\alpha}, \qquad \forall k \geq k_0.$$

Then we get that

$$\int_{\Omega} e^{-f(\omega)}d\mu(\omega) < \infty.$$

Proof Note that we have

$$\int_{\Omega} e^{-f(\omega)}d\mu(\omega) = \int_{\{\omega\in\Omega|f(\omega)<C(\log k)^n\}} e^{-f(\omega)}d\mu(\omega) + \int_{\{\omega\in\Omega|-f(\omega)\geq C(\log k)^n\}} e^{-f(\omega)}d\mu(\omega)$$

$$\leq B_1 + \int_{\mathbb{R}} e^x \mu_f(x)dx$$

$$\leq B_1 + \int_{\mathbb{R}} e^x \mathrm{Exp}\left(-e^{\alpha(\frac{x}{C})^{1/n}}\right) dx < \infty.$$

$\square$

Corollary 5.2.2.3 *Let everything be as before. Then we get that* $e^{-S_{\text{int},\Lambda}[P(\phi)]} \in L^1(\mathcal{S}'(\mathbb{R}^n))$.

Proof Take $f = S_{\text{int},\Lambda}[P(\phi)]$ and $\Omega = \mathcal{S}'(\mathbb{R}^2)$. $\square$

Remark 5.2.2.3 If P is a polynomial of the form $P(x) = \sum\limits_{k=0}^{2n} a_k x^k$ with $a_{2n} > 0$, and $f \in L^1(\mathbb{R}^2) \cap L^2(\mathbb{R}^2)$ with $f \geq 0$, then we can show that $e^{-S_{\text{int},\Lambda}[f,P(\phi)]} \in L^1(\mathcal{S}'(\mathbb{R}^2), \mu)$.

Corollary 5.2.2.4 *Let everything be as before. Then we get that*

$$\frac{e^{-S_{\text{int},\Lambda}[f,P(\phi)]}}{\int_{\mathcal{S}'(\mathbb{R}^2)} e^{-S_{\text{int},\Lambda}[f,P(\phi)]}}$$

is a probability measure on $\mathcal{S}'(\mathbb{R}^2)$.

5.2.3 Theories with Exponential Interaction

Consider the potential

$$V_g^\alpha[\phi] := \int_{\mathbb{R}^2} g(x) : \mathrm{Exp}(\alpha\phi(x)) : dx.$$

We want to define a theory for this type of interaction. Moreover, we want to show that $V_g^\alpha \in L^2(\mathcal{S}'(\mathbb{R}^2), \mu)$ with certain assumption on α and g. Let us first define

$$V_k^\alpha[g] := \int_{\mathbb{R}^2} g(x) : \mathrm{Exp}\left(\alpha\phi\left(\delta_{\frac{1}{k},x}\right)\right) : dx$$

and recall that : $\mathrm{Exp}(\alpha f) := \sum_{k=0}^{\infty} \frac{\alpha^k}{k!} : f^k :$ for $f \in \mathcal{S}(\mathbb{R}^2)$.

Lemma 5.2.3.1 *Let everything be as before. Then we get that* $V_k^\alpha[g] \in L^2(\mathcal{S}'(\mathbb{R}^2), \mu)$, *whenever* $g \in L^1(\mathbb{R}^2) \cap L^2(\mathbb{R}^2)$ *and* $0 \le \alpha^2 \le 4\pi$.

Proof Note that we have

$$\langle : \mathrm{Exp}(\alpha f) :, : \mathrm{Exp}(\alpha g) : \rangle = \mathrm{Exp}(\alpha^2 C(f, g)),$$

and thus we get that

$$\|V_k^\alpha[g]\|^2 = \iint_{\mathbb{R}^2 \times \mathbb{R}^2} g(x)g(y)\mathrm{Exp}\left(\alpha^2 C\left(\delta_{\frac{1}{k},x}, \delta_{\frac{1}{k},x}\right)\right) dxdy$$

$$= \iint_{\mathbb{R}^2 \times \mathbb{R}^2} g(x)g(y)\mathrm{Exp}(\alpha^2 C_k(x, y))dxdy,$$

where $C_k(x, y) = C\left(\delta_{\frac{1}{k},x}, \delta_{\frac{1}{k},y}\right)$. We know that $C_k(x, y) \le C(x, y)$ and that

$$\iint_{\mathbb{R}^2 \times \mathbb{R}^2} g(x)g(y)\mathrm{Exp}(\alpha^2 C(x, y))dxdy < \infty,$$

whenever $0 \le \alpha^2 < 4\pi$, and $g \in L^1(\mathbb{R}^2) \cap L^2(\mathbb{R}^2)$. The latter is true for $\|x - y\| \ge 1$ and $\|x - y\| < 1$ gives $\exp\left(\alpha^2 - \frac{\log\|x-y\|}{2\pi}\right) = \|x - y\| - \frac{\alpha^2}{4\pi}$. Hence $\|V_k^\alpha[g]\|^2 < \infty$. $\qquad\square$

Proposition 5.2.3.1 *Let everything be as before. Then we get that the sequence* $\{V_k^\alpha[g]\}_{k>0}$ *converges in* $L^2(\mathcal{S}'(\mathbb{R}^2), \mu)$.

Proof Define $V_k^\alpha := \sum_{k=0}^{\infty} \frac{\alpha^k}{k!} \int_{\mathbb{R}^2} g(x) : f_k(x)^k : dx$. Then the *Weierstrass M-test* tells us that for any metric space (X, d), Banach space $(W, \|\cdot\|)$ and maps $f_k : X \to W$ with $\|f_k(x)\| \le M_k$ with numbers $M_k > 0$ such that $\sum_{k=0}^{\infty} M_k < \infty$, the sum $\sum_{k=0}^{\infty} f_k(x)$ converges uniformly for x. $\qquad\square$

Exercise 5.2.3.1 Show that there is a constant $C_k > 0$ such that

$$\left\| \frac{\alpha^2}{k!} \int_{\mathbb{R}^2} g(x) : \phi_k(x)^k : dx \right\|^2_{L^2(\mathcal{S}'(\mathbb{R}^2),\mu)} \leq C_k.$$

This implies that V_k^α converges uniformly on X. Thus, we get that

$$\frac{\alpha^k}{k!} \int g(x) : \phi_k(x)^k : dx \longrightarrow \frac{\alpha^k}{k!} \int g(x)(: \phi(x)^k :) dx,$$

where $\phi_k = \phi\left(\delta_{\frac{1}{k},x}\right)$. If $V^\alpha := \lim_{k\to\infty} V_k^\alpha$, we get that

$$V^\alpha = \sum_{k=0}^\infty \frac{\alpha^k}{k!} \int_{\mathbb{R}^2} g(x) : \phi(x)^k : dx = \int_{\mathbb{R}^2} g(x) : \text{Exp}(\alpha\phi(x)) : dx.$$

Thus, we have shown that $V^\alpha \in L^2(\mathcal{S}'(\mathbb{R}^2), \mu)$. Hence, we can observe the following two points:

(1) We get that $V_k^\alpha \geq 0$ for all $k > 0$, whenever $g \geq 0$. This implies that for such g, we get that $V^\alpha \geq 0$ and hence $e^{-V^\alpha} \in L^1(\mathcal{S}'(\mathbb{R}^2))$.
(2) For all $0 \leq \alpha^2 < 4\pi$ and $g \geq 0$ with $g \in L^1(\mathbb{R}^2) \cap L^2(\mathbb{R}^2)$, we have shown that $e^{-V^\alpha[g]} \in L^1(\mathcal{S}'(\mathbb{R}^2), \mu)$. Let ν be a measure on $[-\alpha, \alpha] \subset \mathbb{R}$. Then we get that

$$\int_{[-\alpha,\alpha]} e^{-V^{\alpha'}[g]} d\nu(\alpha') \in L^1(\mathcal{S}'(\mathbb{R}^2), \mu).$$

5.2.4 The Osterwalder–Schrader Axioms

The *Osterwalder–Schrader axioms* (Osterwalder and Schrader 1975) are axioms for the description of a Euclidean quantum field theory for a Hermitian scalar field. Similarly to the setting of quantum mechanics, we want to understand how such a theory is fully described through a set of axioms. Let μ be a Borel measure on $\mathcal{S}'(\mathbb{R}^n)$.

5.2.4.1 Analyticity (OS0)

Let $f_1, ..., f_k \in \mathcal{S}'(\mathbb{R}^n)$ and define a function

$$\hat{\mu}(f_1, ..., f_k) \colon \mathbb{C}^k \longrightarrow \mathbb{C},$$

$$(z_1, ..., z_k) \longmapsto \hat{\mu}(f_1, ..., f_k)(z_1, ..., z_k) := \hat{\mu}\left(\sum_{j=1}^{k} z_j f_j\right),$$

where $\hat{\mu}$ is the characteristic functional of μ, i.e., we have

$$\hat{\mu}(f) = \int_{\mathcal{S}'(\mathbb{R}^n)} e^{i\phi(f)} d\mu(\phi).$$

Definition 5.2.4.1 (Analyticity). We say that μ is *analytic* or μ has *analyticity* if $\hat{\mu}(f_1, ..., f_k)$ is an entire function on $\mathbb{C}^k$ for all $f_1, ..., f_k \in \mathcal{S}(\mathbb{R}^n)$ and $k \in \mathbb{N}$. This means that μ decays faster than any exponential map.

Remark 5.2.4.1 An immediate consequence is that $\int_{\mathcal{S}'(\mathbb{R}^n)} \phi(f) d\mu(\phi) < \infty$ for all k and for all $f \in \mathcal{S}(\mathbb{R}^n)$. Then we get that

$$\hat{\mu}(if) = \int_{\mathcal{S}'(\mathbb{R}^n)} e^{-\phi(f)} d\mu(\phi) < \infty,$$

and

$$\hat{\mu}(-if) = \int_{\mathcal{S}'(\mathbb{R}^n)} e^{\phi(f)} d\mu(\phi) < \infty,$$

if μ is analytic.

Example 5.2.4.1 Let μ be the Gaussian measure on $\mathcal{S}'(\mathbb{R}^n)$, whose covariance is given by $(\Delta + m^2)^{-1}$, and let $C(f, g) = \iint_{\mathbb{R}^n \times \mathbb{R}^n} f(x)C(x, y)g(y)dxdy = \langle f, (\Delta + m^2)^{-1}g \rangle_{C^2(\mathbb{R}^n)}$. We claim that μ has analyticity. We prove this via an example. Consider

$$\hat{\mu}(z_1 f_1 + z_2 f_2) = \int_{\mathcal{S}'(\mathbb{R}^n)} e^{i\phi(z_1 f_1 + z_2 f_2)} d\mu(\phi) = e^{-\frac{1}{2}(2z_1 z_2 C(f_1, f_2) + z_1^2 C(f_1, f_1) + z_2^2 C(f_2, f_2))},$$

which is obviously an entire function, and $\hat{\mu}(z_1 f_1 + z_2 f_2) = \hat{\mu}(f_1, f_2)(z_1, z_2)$, which is analytic.

Proposition 5.2.4.1 *Let ν be any Gaussian measure on $\mathcal{S}'(\mathbb{R}^n)$. Then ν has analyticity.*

5.2.4.2 Euclidean Invariance (OS1)

Let $E(n)$ be the Euclidean group of $\mathbb{R}^n$, i.e., the group generated by rotations, reflections and translations in $\mathbb{R}^n$. Let $R \in O(n)$ and $a \in \mathbb{R}^n$ and let $T(a, R) \in E(n)$ be defined by $T(a, R)(x) := Rx + a$. Note that $E(n)$ acts on $\mathcal{S}(\mathbb{R}^n)$ by

$(T(a, R)f)(x) = f(T(a, R)^{-1}x)$. The group $E(n)$ acts also on $S'(\mathbb{R}^n)$ by $(T(a, R)\phi)(f) = \phi(T(a, R)f)$.

Definition 5.2.4.2 (Euclidean invariance I). We say that μ is *Euclidean invariant* if

$$T(a, R)_*\mu = \mu, \quad \forall T(a, R) \in E(n).$$

Lemma 5.2.4.1 *The measure μ is Euclidean invariant if and only if*

$$\hat{\mu}(f) = \hat{\mu}(T(a, R)f), \quad \forall f \in S(\mathbb{R}^n).$$

Definition 5.2.4.3 (Euclidean invariance II). Let ν be a Gaussian measure on $S'(\mathbb{R}^n)$ whose covariance is given by a map $C_\nu \colon S(\mathbb{R}^n) \times S(\mathbb{R}^n) \to \mathbb{R}$. We say that C_ν is *Euclidean invariant* if the covariance $\mathrm{Cov}(T(a, R)f, T(a, R)g) = C_\nu(f, g)$ for all $T(a, R) \in E(n)$ and $f, g \in S(\mathbb{R}^n)$.

Lemma 5.2.4.2 *Let ν be a Gaussian measure. Then ν is Euclidean invariant if and only if C_ν is Euclidean invariant.*

Example 5.2.4.2 Let μ be the Gaussian measure on $S'(\mathbb{R}^n)$ with covariance $(\Delta + m^2)^{-1}$. Then μ is Euclidean invariant. In particular, we have

$$C(x, y) = \frac{1}{(2\pi)^n} \int_{\mathbb{R}^n} \frac{e^{i\xi_1 \|x-y\|}}{\xi^2 + m^2} d\xi.$$

Next, we want to construct a Hilbert space $\mathcal{E} := L^2(S'(\mathbb{R}^n), \mu)$.

5.2.4.3 Reflection Positivity (OS2)

Let $f_1, ..., f_k \in S(\mathbb{R}^n)$, such that $\mathrm{supp}(f_i) \subseteq \mathbb{R}^n_+$ for all $i = 1, ..., k$. Write $\mathbb{R}^n = \mathbb{R}^{n-1} \times \mathbb{R}$ and $\mathbb{R}^n_+ = \mathbb{R}^{n-1} \times (0, \infty)$.

Definition 5.2.4.4 (Reflection positivity I). We say that the measure μ has *reflection positivity* if for all $z_1, ..., z_k \in \mathbb{C}$ we have $\sum_{i,j} \bar{z}_i \hat{\mu}(f_i \cdot \mathrm{Im}(\theta) \cdot f_j)z_j \geq 0$, where $\theta(x, t) = (x, -t)$ for all $k \in \mathbb{N}$, $f_1, ..., f_k \in S(\mathbb{R}^n)$ with $\mathrm{supp}(f_j) \subseteq \mathbb{R}^n_+$.

Assume that ν is a Gaussian measure and let C_ν be its covariance.

Definition 5.2.4.5 (Reflection Positivity II). We say that C_ν has *reflection positivity* if

$$C_\nu(f, \theta \circ f) = \iint_{\mathbb{R} \times \mathbb{R}} \iint_{\mathbb{R}^{n-1} \times \mathbb{R}^{n-1}} f(x, t)f(y, -s)C((x, t), (y, s)) dx dy dt ds \geq 0$$

for all $f \in S(\mathbb{R}^n)$ such that $\mathrm{supp}(f) \subseteq \mathbb{R}^n_+$.

Exercise 5.2.4.1 Let ν be a Gaussian measure on $\mathcal{S}'(\mathbb{R}^n)$. Show that ν has reflection positivity if and only if C_ν has reflection positivity.

Example 5.2.4.3 Let μ be the Gaussian measure with covariance $(\Delta + m^2)^{-1}$. Then μ has reflection positivity.

Proof We will show that $C(f, g) = \langle f, (\Delta + m^2)^{-1} g \rangle_{L^1(\mathbb{R}^n)}$ is reflection positive. To do this, let $f \in \mathcal{S}(\mathbb{R}^n)$ with $\operatorname{supp}(f) \subseteq \mathbb{R}^n_+$, and $x = (\bar{x}, t) \in \mathbb{R}^{n-1} \times \mathbb{R}$. Then we get

$$
\begin{aligned}
C(f, g) &= \iint_{\mathbb{R}^n \times \mathbb{R}^n} f(x) C(x, y) g(y) \mathrm{d}x \mathrm{d}y \\
&= \frac{1}{(2\pi)^n} \iint_{\mathbb{R}^n \times \mathbb{R}^n} f(x) \left(\int_{\mathbb{R}^n} \frac{e^{i\xi(x-y)}}{\xi^2 + m^2} \mathrm{d}\xi \right) g(y) \mathrm{d}x \mathrm{d}y \\
&= \int_{\mathbb{R}^n} \frac{\hat{f}(\xi) \hat{g}(\xi)}{\xi^2 + m^2} \mathrm{d}\xi.
\end{aligned}
$$

Moreover, we have $C(f, \theta \circ f) = \int_{\mathbb{R}^n} \frac{\overline{\theta(f(\xi))} \hat{f}(\xi)}{\xi^2 + m^2} \mathrm{d}\xi$, which we want to be positive. Furthermore, we have

$$
\hat{f}(\vec{\xi}, i\xi_n) = \frac{1}{(2\pi)^{n/2}} \int_0^\infty \left(\int_{\mathbb{R}^{n-1}} f(x, t) e^{i\langle \vec{\xi}, x \rangle - \xi_n t} \mathrm{d}\vec{\xi} \right) \mathrm{d}t,
$$

where $\vec{\xi} = (\xi_1, ..., \xi_{n-1}) \in \mathbb{R}^{n-1}$. Similarly, we have

$$
\widehat{\theta \circ f}(\vec{\xi}, i\xi_n) = \frac{1}{(2\pi)^{n/2}} \int_0^\infty \left(\int_{\mathbb{R}^{n-1}} \overline{f(x, t)} e^{-i\langle \vec{\xi}, x \rangle - \xi_n t} \mathrm{d}\vec{\xi} \right) \mathrm{d}t.
$$

Thus, we get that

$$
\hat{f}(\vec{\xi}, i\xi_n) = \overline{\widehat{\theta f}(\vec{\xi}, i\xi_n)}.
$$

Using these relations, we can show that

$$
C(f, \theta \circ f) = \int_{\mathbb{R}^{n-1}} \frac{|\hat{f}(\vec{\xi}, i\mu(\vec{\xi}))|^2}{\sqrt{\vec{\xi}^2 + m^2}} \mathrm{d}\vec{\xi} \geq 0,
$$

where $\mu(\vec{\xi}) = \sqrt{\vec{\xi}^2 + m^2}$. $\qquad\square$

If we consider the measure space $\mathcal{E} := L^2(\mathcal{S}'(\mathbb{R}^n), \nu)$, we assume that ν has analyticity and Euclidean invariance. Moreover, consider the set

$$
\mathcal{A} := \left\{ A[\phi] = \sum_{j=1}^k c_j e^{i\phi(f_j)} \, \Big| \, c_j \in \mathbb{C}, k \in \mathbb{N} \right\}.
$$

In fact, we have $\mathcal{A} \subseteq \mathcal{E}$, because of analyticity, and since $\mathcal{A}$ is an algebra. Moreover, define the set $\mathcal{E}_+ := \{A(\phi) \in \mathcal{A} \mid \mathrm{supp}(f_i) \subseteq \mathbb{R}^n_+\} \subseteq \mathcal{E}$. We also define a bilinear form b on $\mathcal{E}_+$ by $b(A, B) := \int \overline{\theta \circ A} B d\nu(\phi)$.

Exercise 5.2.4.2 Show that the measure ν is reflection positive if and only if for all $A \in \mathcal{E}_+$ we get $b(A, A) \geq 0$.

Definition 5.2.4.6 (Physical Hilbert space). Define the set $N := \{A \in \mathcal{E}_+ \mid b(A, A) = 0\}$, and let $\mathcal{H}$ be the completion of the space $\mathcal{E}_+/N$. Then the space $\mathcal{H}$ is called the *physical Hilbert space*.

We can observe that if we have $T : \mathcal{E} \to \mathcal{E}$ such that $T(\mathcal{E}_+) \subseteq \mathcal{E}_+$ and $T(N) \subseteq N$, then T induces a map $T(t) : \mathcal{H} \to \mathcal{H}$, where $T(t)(\vec{x}, s) = (\vec{x}, s + t)$ for $t \geq 0$. We know that $T(t)$ acts on $\mathcal{E}$ unitarily.

Lemma 5.2.4.3 *We have $T(t)(\mathcal{E}_+) \subseteq \mathcal{E}_+$ and $T(t)(N) \subseteq N$.*

Proof The first part is obvious. For the second part, observe that $\theta \circ T(t) = T(-t) \circ \theta$. Let $A \in N$. Then

$$\langle T(t)(A), (\theta \circ T(t))(A) \rangle_{\mathcal{E}} = \langle T(t)(A), (T(-t) \circ \theta)(A) \rangle_{\mathcal{E}} = \langle T(2t)(A), \theta(A) \rangle_{\mathcal{E}}$$
$$= b(T(2t)(A), A) \leq \underbrace{b(A, A)^{1/2}}_{=0} b(T(2t)(A), T(2t)(A))^{1/2} = 0,$$

which implies that $T(t)(A) \in N$ because of reflection positivity. $\qquad\square$

One can also check that the map $T(t) : \mathcal{H} \to \mathcal{H}$ is a semi-group for $t \geq 0$.

Lemma 5.2.4.4 *We have $\|T(t)\|_{\mathcal{H}} \leq 1$, for $t \geq 0$. Moreover, the map $t \mapsto T(t)$ is strongly continuous.*

Corollary 5.2.4.1 *If H is a positive self-adjoint operator on $\mathcal{H}$, we get that $T(t) = e^{-tH}$. Moreover, we have $H(1) = 0$.*

Example 5.2.4.4 (Free massive scalar field theory). Consider a measure μ and the Green's function $C(x, y)$. We want to know whether we can find an "explicit" representation of $\mathcal{H}$ in terms of time zero hyper-surfaces in $\mathbb{R}^{n-1}$. We can indeed write $\mathcal{H} \cong L^2(\mathcal{S}'(\mathbb{R}^{n-1}), \nu) \subseteq \Gamma(H^{-1}(\mathbb{R}^n)) = L^2(\mathcal{S}'(\mathbb{R}^n), \mu)$, where ν is a Gaussian measure.

Let $f \in \mathcal{S}(\mathbb{R}^{n-1})$ and define the map $j_0(f) := f \otimes \delta_0$, where $(f \otimes \delta_0)(\vec{x}, t) = f(\vec{x}) \cdot \delta_0(t)$. We then claim that $f \otimes \delta_0 \in H^{-1}(\mathbb{R}^n_+)$. Indeed, we have $\widehat{f \otimes \delta_0}(\vec{\xi}, \xi_n) = \hat{f}(\vec{\xi})$, and we know

$$\langle f \otimes \delta_0, C(f \otimes \delta_0) \rangle = \frac{1}{2\pi} \int_{\mathbb{R}^n} \frac{|\hat{f}(\vec{\xi})|^2}{\xi^2 + m^2} d\xi$$

$$= \frac{1}{2\pi} \int_{\mathbb{R}^{n-1}} |\hat{f}(\vec{\xi})|^2 \left(\int_{\mathbb{R}} \frac{1}{\xi^2 + m^2} d\xi_n \right) d\vec{\xi} = \frac{1}{2} \int_{\mathbb{R}^{n-1}} \frac{|\hat{f}(\vec{\xi})|^2}{\sqrt{\vec{\xi}^2 + m^2}} d\vec{\xi}.$$

Thus, we get that $f \otimes \delta_0 \in H^{-1}(\mathbb{R}^n)$. Moreover, we get that

$$\langle f \otimes \delta_0, C(f \otimes \delta_0) \rangle_{L^2(\mathbb{R}^n)} = \frac{1}{2} \left\langle f, \left(\sqrt{\Delta_{\mathbb{R}^{n-1}} + m^2} \right)^{-1} f \right\rangle_{L^2(\mathbb{R}^{n-1})}.$$

If we define $B(f, g) := \frac{1}{2} \int_{\mathbb{R}^{n-1}} f \left(\Delta_{\mathbb{R}^{n-1}} + m^2 \right)^{-1/2} g dx$, we can see that j_0 defines an isometry $\overline{\mathcal{S}(\mathbb{R}^n)}^{B(\nu)} \to H^{-1}(\mathbb{R}^n)$, where $\overline{\mathcal{S}(\mathbb{R}^n)}^{B(\nu)}$ is the completion of $\mathcal{S}(\mathbb{R}^n)$ with respect to B.

Lemma 5.2.4.5 *For $t \in \mathbb{R}$, we define $j_t(f) := f \otimes \delta_t$, with $f \in \mathcal{S}(\mathbb{R}^{n-1})$. Then for $t \geq s$, we get that*

$$\langle j_t(f), j_s(g) \rangle_{L_2(\mathbb{R}^n)} = \frac{1}{2} \left\langle f, \left(\Delta_{\mathbb{R}^n} + m^2 \right)^{-1/2} e^{-(t-s)\sqrt{\Delta + m^2}} g \right\rangle_{L^2(\mathbb{R}^n)}.$$

Let ν be the Gaussian measure on $\mathcal{S}'(\mathbb{R}^{n-1})$ whose covariance is B and denote by

$$H^{-1/2}(\mathbb{R}^{n-1}) := \overline{\mathcal{S}(\mathbb{R}^n)}^{B(\nu)}.$$

Then we know that $L^2(\mathcal{S}'(\mathbb{R}^{n-1}), \nu) \cong \Gamma(H^{-1/2}(\mathbb{R}^{n-1}))$. Given an operator A on $\mathcal{H}$, one can define an operator $d\Gamma(A)$ on $\Gamma(\mathcal{H})$ as follows. On $\text{Sym}^n(\mathcal{H})$ we get

$$d\Gamma(A) = A \otimes I \otimes \cdots \otimes I + I \otimes A \otimes I \otimes \cdots \otimes I + ...,$$

and on $\text{Sym}^0(\mathcal{H}) = \mathbb{C}$ we get $d\Gamma(A) = 0$. Here we have denoted by I the identity operator on $\mathcal{H}$. Moreover, if we identify $\mathcal{H}$ with $L^2(\mathcal{S}'(\mathbb{R}^{n-1}), \nu) \cong \Gamma(H^{-1/2}(\mathbb{R}^{n-1}))$, we get that

$$d\Gamma(\sqrt{\Delta_{\mathbb{R}^n} + m^2}) = 0.$$

5.3 QFT as an Operator-Valued Distribution

The motivation of this section is to get a better understanding of relativistic quantum mechanics. Recall that the data for a quantum mechanical system is given by:

- a *Hilbert space* of states $\mathcal{H}$ (e.g., $L^2(\mathbb{R}^n)$);
- *observables*, which are represented by self-adjoint operators on $\mathcal{H}$;

- *symmetries*, which are unitary representations on $\mathcal{H}$, and 1-parameter group of symmetries, lead to specific observables (e.g., time-translation leads to the Hamiltonian $\hat{H}$ of the system);
- some *dynamics*, which are controlled by the *Schrödinger equation* $i\hbar\frac{\partial\psi}{\partial t} = \hat{H}\psi$.

5.3.1 Relativistic Quantum Mechanics

In relativistic quantum mechanics we want to have a unitary representation of the *Poincaré group* $\mathcal{P}$, which is the group of all *space-time symmetries*. Recall that the Minkowski space-time is given by $\mathbb{M}^n := \mathbb{R}^{1,n-1}$, where we can have coordinates in *position space* (such as $x = (t, \vec{x})$) or in *momentum space* (such as $(\xi = (\xi_0, \vec{\xi}))$). Denote by $\mathcal{L}$ the *Lorentz group*, which is the set of all linear isometries of $\mathbb{M}^n$, i.e., a set of the form

$$\left\{ (\Lambda_{ij})_{\substack{1 \le i \le n \\ 1 \le j \le n}} \in \mathrm{Mat}_{n\times n}(\mathbb{M}^n) \;\middle|\; \Lambda^T g \Lambda = g, \; \forall g \in \mathbb{M}^n \right\},$$

thus for $\Lambda \in \mathcal{L}$ we have $\det \Lambda \in \{\pm 1\}$. Moreover, we can write $\mathcal{L}$ as a union of subspaces:

$$\mathcal{L} = \mathcal{L}_+^\uparrow \cup \mathcal{L}_1^\uparrow \cup \mathcal{L}_+^\downarrow \cup \mathcal{L}_-^\downarrow,$$

where the label $\uparrow$ ($\downarrow$) means the determinant is $+1$ (-1), and the label $+$ ($-$) means $\Lambda_{00} > 0$ (< 0). Note that the identity $I \in \mathcal{L}_+^\uparrow$, which we call the restricted Lorentz group. We can then define the *Poincaré group* by

$$\mathcal{P} := \left\{ T(\Lambda, a) \mid T(\Lambda, a)(x) = \Lambda x + a, \; \forall x \in \mathbb{R}^n, \; \forall \Lambda \in \mathcal{L}, \; \forall a \in \mathbb{R}^n \right\}.$$

Thus, we can write the Poincaré group as a semi-direct product $\mathcal{P} = \mathcal{L} \ltimes \mathbb{R}^n$. We can write $\mathcal{P}$ as the union of subspaces in the same way as for $\mathcal{L}$. We call $\mathcal{P}_+^\uparrow$ the restricted Poincaré group. We want to have a projective unitary representation of $\mathcal{P}_+^\uparrow$.

5.3.1.1 Bergmann's Construction ($n = 4$)

In this construction, the projective unitary representation of $\mathcal{P}_+^\uparrow$ comes from the unitary representation of $\widetilde{\mathcal{P}_+^\uparrow}$, which is the universal cover of $\mathcal{P}_+^\uparrow$. In fact, we have that $SL(2, \mathbb{C})$ is the universal cover of $\mathcal{L}_+^\uparrow$. Hence, in this case, we get that

$$\widetilde{\mathcal{P}_+^\uparrow} = SL(2, \mathbb{C}) \ltimes \mathbb{R}^4.$$

5.3.1.2 Wigner's Construction

Let us consider $p \in \mathbb{R}^4$, and let H_p be the stabilizer of p by the action of $SL(2, \mathbb{C})$. Moreover, take a unitary irreducible representation $\mathcal{H}_{\sigma,p}$ of H_p. One can then use the so-called *Mackey machinery*. Choose a G-invariant measure on G/H_p and define the Hilbert space $\mathcal{H}$ to be the $\mathcal{H}_{\sigma,p}$-valued functions on G/H_p and use the invariant measure to define an inner product.

Proposition 5.3.1.1 (Wigner). *The space $\mathcal{H}$ is an irreducible unitary representation of G. Moreover, all irreducible unitary representations of G appear in this way.*

Remark 5.3.1.1 The stabilizer group H_p can have $(2s + 1)$-dimensional irreducible representations. Here $s \in \{0, \frac{1}{2}, 1\}$ represents the "spin" of the particle.

Assume $s = 0$. We will start with the trivial representation, which is a 1-dimensional representation of H_p. Consider the sets

$$X_m^+ := \left\{ \xi = (\xi_0, \vec{\xi}), \in \mathbb{R}^4 \,\middle|\, \xi^2 - m^2 = 0, \, \xi_0 > 0 \right\},$$

$$X_m^- := \left\{ \xi = (\xi_0, \vec{\xi}), \in \mathbb{R}^4 \,\middle|\, \xi^2 - m^2 = 0, \, \xi_0 < 0 \right\},$$

and define $X_m := X_m^+ \cup X_m^-$ and $X := \bigcup_{m \geq 0} X_m$. Consider the point $p = (m, 0, 0, 0) \in X$. Then we get that $G/H_p = X_m^+$. Now we want to construct an invariant measure on X_m^+. For this, let f be a positive function on $(0, \infty)$. Then we get that $f(\xi^2)\mathrm{d}\xi$ is an invariant measure on X. We would like to have an invariant measure of the form $\delta(\xi^2 - m^2)\mathrm{d}\xi$. We can define a map

$$\phi \colon (0, \infty) \times \mathbb{R}^3 \longrightarrow X_m^+,$$

$$(y, \vec{\xi}) \longmapsto \left(\sqrt{y + |\vec{\xi}|^2}, \vec{\xi} \right).$$

Then we can see that $\phi^*(f(\xi^2)\mathrm{d}\xi) = \frac{f(y)\mathrm{d}y\mathrm{d}\vec{\xi}}{\sqrt{y^2 + |\vec{\xi}|^2}}$. We want to have the pushforward of $\delta_m := \frac{\mathrm{d}y\mathrm{d}\vec{\xi}}{\sqrt{y^2 + |\vec{\xi}|^2}}$ to be our measure on X_m^+. More precisely, define a map

$$\alpha \colon \mathbb{R}^3 \longrightarrow X_m^+,$$

$$\vec{\xi} \longmapsto \alpha(\vec{\xi}) := \left(\sqrt{m^2 + |\vec{\xi}|^2}, \vec{\xi} \right).$$

As an invariant measure, we want to get the pushforward of $\frac{1}{2\sqrt{m^2 + |\vec{\xi}|^2}}\mathrm{d}\vec{\xi}$ on $\mathbb{R}^3$ to X_m^+. Wigner's theorem tells us that $\mathcal{H} = L^2(X_m^+, \mu_m) \cong L^2(\mathbb{R}^3, \nu)$, where $\frac{\mathrm{d}\nu}{\mathrm{d}\vec{\xi}} = \frac{1}{2\sqrt{m^2 + |\vec{\xi}|^2}}$. The position operator is then given by $\frac{1}{2}(\Delta + m^2)^{-1/2}$. One can summarize the result by saying that the Hilbert space for *spin zero particles* is given by $L^2(\mathbb{R}^3, \nu) \cong H^{-1/2}(\mathbb{R}^3)$.

5.3.2 Garding–Wightman Formulation of QFT

The *Garding–Wightman axioms* (Wightman and Garding 1964) are another set of axioms in constructive quantum field theory in order to give a rigorous construction basis for the theory of quantum fields. The axioms are given by the following points.

(GW1) There is a Hilbert space $\mathcal{H}$, a vacuum state $\Omega \in \mathcal{H}$, and a unitary representation $\mathcal{P}_+^\uparrow$ on $\mathcal{H}$.

(GW2) There is a field operator Φ as a map from $\mathcal{S}(\mathbb{R}^4)$ to operators on $\mathcal{H}$ together with a dense subspace D of $\mathcal{H}$ such that:

    (a) $\Omega \in D$,
    (b) $D \subseteq D(\Phi(f))$ for all f,
    (c) the map $f \mapsto \Phi(f)|_D$ is linear,
    (d) for all $\Omega_1, \Omega_2 \in D$, the map $f \mapsto \langle \Phi(f)\Omega_1, \Omega_2 \rangle$ is a Schwartz distribution (regularity),
    (e) $\Phi(f)^* = \Phi(\bar{f})$.

(GW3) (Covariance) We have that:

    (a) $U(\Lambda, a)D \subseteq D$, where $U(\Lambda, a)$ is the unitary representation of $T(\Lambda, a) \in \mathcal{P}_+^\uparrow$ on $\mathcal{H}$,
    (b) $U(\Lambda, a) \cdot \Phi(f) \cdot U(\Lambda, a)^{-1} = \Phi(T(\Lambda, a)(f))$.

(GW4) (Spectrum) Since $\mathbb{R}^4$ acts unitarily on $\mathcal{H}$ via $U(\Lambda, a)$, we can take $P_1, ..., P_4$ to be the infinitesimal generators of this action. One can show that $P_1, ..., P_4$ are essentially self-adjoint. The axiom is then formulated as follows: The joint spectrum of $(P_1, ..., P_4)$ lies in $X^+ = \{\xi \in \mathbb{R}^4 \mid \xi^2 \geq 0, \xi_0 > 0\}$, where physically we have $\xi^2 = E^2 - |\vec{p}|^2$ with energy $E > 0$ and momentum $\vec{p}$.

(GW5) (Locality) If f and g have space-like disjoint support, then $[\Phi(f), \Phi(g)] = 0$.

Remark 5.3.2.1 By the Garding–Wightman axioms, one can show that the vacuum is unique. Indeed, if $U(\Lambda, a)\Omega' = \Omega'$ for all $T(\Lambda, a) \in \mathcal{P}_+^\uparrow$, then $\Omega' = c \cdot \Omega$, where $c \in \mathbb{C}$.

Given $f_1, ..., f_k \in \mathcal{S}(\mathbb{R}^4)$, we can define the map $(f_1, ..., f_k) \mapsto \langle \Omega, \Phi(f_1) \cdots \Phi(f_k)\Omega \rangle$, which, by (GW2), is a distribution in $\mathcal{S}'(\mathbb{R}^4)$, i.e., we have

$$\mathcal{W}(f_1, ..., f_k) := \langle \Omega, \Phi(f_1) \cdots \Phi(f_k)\Omega \rangle = W_k(f_1 \otimes \cdots \otimes f_k).$$

Definition 5.3.2.1 (Wightman distribution). The distributions $W_k \in \mathcal{S}'(\mathbb{R}^4)$ are called *Wightman distribution*.

5.3.2.1 The Wightman Axioms

We are now able to formulate the *Wightman axioms*:

(W1) W_k are $\mathcal{P}_+^\uparrow$-invariant,

(W2) If $f_1 \in \mathcal{S}(\mathbb{R}^4)$, ..., $f_k \in \mathcal{S}(\mathbb{R}^{4k})$, then $\sum\limits_{i,j=0}^{k} W_{i+j}(\bar{f}_i \otimes f_j) \geq 0$.

(W3) (Locality) We have that

$$W_k(f_1 \otimes \cdots \otimes f_j \otimes f_{j+1} \otimes \cdots \otimes f_k) = W_k(f_1 \otimes \cdots \otimes f_{j+1} \otimes f_j \otimes \cdots \otimes f_k),$$

whenever f_j and f_{j+1} are space-like separated.

(W4) (Spectral condition) For all $n > 0$ there is a tempered distribution[5] $T_n \in \mathcal{S}'(\mathbb{R}^{4n-4})$, such that for all $f \in \mathcal{S}(\mathbb{R}^{4n})$ with the property that its Fourier transform has support in $\underbrace{X^+ \times \cdots \times X^+}_{n-1}$, we get that $W_n(f) = T_n(\hat{f})$, where $\hat{f} \in \mathcal{S}(\mathbb{R}^{4n-4})$ is defined through

$$\hat{f}(x_1, ..., x_{n-1}) := f(0, x_1 + x_2, ..., x_1 + \cdots + x_{n-1}), \qquad \forall x \in \mathbb{R}^4.$$

(W5) (Cluster property) If $a \in \mathbb{R}^4$ is space-like, then for all $0 \leq i \leq n$ we have that

$$\lim_{\lambda \to 0} W_n \otimes T_{\lambda a,i} = W_i \otimes W_{n-i},$$

where $T_{a,i} \colon \mathcal{S}(\mathbb{R}^{4n}) \to \mathcal{S}(\mathbb{R}^{4n})$ is the translation operator defined through $T_{a,i} f(x_1, ..., x_n) = f(x_1, ..., x_i, x_i - a, ..., x_n - a)$.

Theorem 5.3.2.1 (Wightman reconstruction theorem). *If we have distributions* $(W_k)_k$ *satisfying (W1)–(W6), then there is a "unique" GW quantum field theory* $(\mathcal{H}, U, \Omega, \Phi, D)$, *whose Wightman distributions are* W_k, *i.e., we have*

$$W_k(f_1 \otimes \cdots \otimes f_k) = \langle \Omega, \Phi(f_1) \cdots \Phi(f_k)\Omega \rangle.$$

5.4 Outlook

5.4.1 Generalization and Gauge Theories

In this book we have discussed the constructive approach to Euclidean quantum field theory and the approach of Feynman's functional integral method. The functional

---

[5] A tempered distribution can be formally defined as a continuous linear functional on the Schwartz space $\mathcal{S}(\mathbb{R}^n)$ of smooth functions with rapidly decreasing derivatives.

integral approach, of course, needs to be understood in many different cases in order to be useful for other field theories of interest. For many such classical field theories, the space of fields is actually given by some complicated infinite-dimensional manifold,[6] where it is actually not possible to construct a measure $\mathscr{D}$ for the functional integral. It is not immediately clear what "quantization" in this setting is supposed to be when formally applying this construction. The way out of this is to consider a *perturbative expansion* of the functional integral and define an object of the form

$$\int_{\phi \in F} e^{\frac{i}{\hbar} S[\phi]} \mathscr{D}[\phi]$$

perturbatively in terms of a *formal power series* in $\hbar$. Here we have denoted by F the space of fields for a given classical field theory and by S the classical action. In the case we have discussed before, we had $F = C^\infty(\mathbb{R}^n)$ and we have Wick rotated the theory in order to have it Euclidean. The method of expanding such functional integrals into formal power series is given by *stationary phase expansion*. Another important thing is to understand what happens if the theory of interest has *symmetries*, i.e., if we have a local theory of the form $S[\phi] = \int_M \mathcal{L}(\phi, \partial\phi, \ldots, \partial^N \phi)$, where M is some manifold and $\mathcal{L}$ denotes the *Lagrangian density* for the theory with $N \in \mathbb{N}$, there is some Lie group G that, when acting on the Lagrangian, leaves the theory invariant. These types of theories are usually called *gauge theories*. An important example for a gauge theory is electrodynamics, since when replacing the vector potential $\mathbf{A}$ with $\mathbf{A} + \nabla\lambda$, for some function λ, we still get the same magnetic field $\mathbf{B} = \nabla \times \mathbf{A} = \nabla \times (\mathbf{A} + \nabla\lambda)$. The problem that appears within such theories is when we want to perform the functional integral quantization. When applying the method of stationary phase expansion, there is a term of the form $\frac{1}{\sqrt{|\det(\partial^2 S(\phi_0))|}}$, where ϕ_0 is a critical point of the action S, i.e., a solution to the Euler–Lagrange equations $\delta S = 0$ and one sums over all the critical points. Moreover, we have denoted by $\partial^2 S$ the Hessian of S. This means that we need to make sure that all the critical points of S are isolated. Now if S is a gauge theory, the critical points will never be isolated, because they will always appear in G-orbits. This was a big problem for modern quantum field theory, especially since most of the theories of interest, such as e.g., Yang–Mills theory, are gauge theories. A powerful method, suited for Yang–Mills theories and similar kinds, is the method of *ghost fields* which was provided by Faddeev and Popov (Faddeev and Popov 1967). This method rewrites the functional integral of interest into another one by using some non-physical fields—the ghost fields—such that critical points of the new action all become isolated. At the end, the ghost fields will be integrated out and one ends up again with the original field configuration. In fact, several generalizations of this construction have been studied. A *cohomological approach* was considered by Becchi, Rouet and Stora (Becchi et al. 1976) and independently by Tyutin (Tyutin 1975). A generalization of their approach to the *symplectic* case was considered later by Batalin and Vilkovisky (Batalin and

---

[6] Usually, we want them to be locally modelled by a Fréchet space.

Vilkovisky 1981) in order to substantially increase the amount of theories that can be treated by this formalism.

5.4.2 TQFTs and the Functorial Approach

Another approach to quantum field theory is through the *functorial* picture. For simplification we only want to consider the case where the classical theory is *topological*, i.e., the case where the action functional is invariant with respect to reparametrization. In other words, there is no metric on the underlying manifold. This is due to Atiyah (Atiyah 1988), who gave an axiomatic construction. One usually defines a *topological quantum field theory (TQFT)* as a functor

$$Z: \mathbf{Cob}_n \longrightarrow \mathbf{Vect}_{\mathbb{C}},$$

where $\mathbf{Cob}_n$ denotes the category of *n-cobordisms* and $\mathbf{Vect}_{\mathbb{C}}$ denotes the category of $\mathbb{C}$-vector spaces. The objects in the category $\mathbf{Cob}_n$ are given by closed $(n-1)$-manifolds and the morphisms are given by the bounding n-manifolds between them. Hence, to each object Σ we assign a vector space $\mathcal{H} := Z(\Sigma)$. Moreover, each object Σ is equipped with a certain coloring which is related to an "orientation". We have two different colors assigned to an object, called "in" and "out". To Σ^{in} we assign a vector space $\mathcal{H}$ and to Σ^{out} we assign the dual space $\mathcal{H}^*$. To each morphism n-manifold M with boundary $\partial M = \bigsqcup_i \Sigma_i^{\text{in}} \sqcup \bigsqcup_j \Sigma_j^{\text{out}}$ we assign a homomorphism $Z_M \in \bigotimes_i \mathcal{H}_i \otimes \bigotimes_j \mathcal{H}_j^*$, i.e., a map $Z_M: \bigotimes_j \mathcal{H}_j \to \bigotimes_i \mathcal{H}_i$. In fact, the functor Z is *symmetric monoidal* for the monoidal structure on $\mathbf{Cob}_n$ given by disjoint union $\sqcup$ and the one on $\mathbf{Vect}_{\mathbb{C}}$ given by the tensor product $\otimes$. This construction can be generalized to the case when we have some geometrical structure, e.g., a metric (Segal 1988). This approach gives the advantage that we do not need to make sense of some ill-defined measure. In fact, it is the approach that is considered the most acceptable for mathematicians. In particular, it relies also on the theory of locality which allows us to cut morphisms and glue objects. The gluing of objects would correspond to the dual pairing of vector spaces, so we are only allowed to glue "in"-colored to "out"-colored objects. However, there is no action functional in this picture. If the manifold M is closed it is considered as a morphism from the empty set to the empty set. Since Z is a symmetric monoidal functor, it maps the empty set to the field of complex numbers. This means that $Z_M \in \mathbb{C}$. This is indeed the case when we compute it as a functional integral

$$Z_M = \int_{\phi \in F_M} e^{\frac{i}{\hbar} S[\phi]} \mathscr{D}[\phi] \in \mathbb{C}.$$

In the case where the manifold M has boundary, this object would be an element of the Hilbert space assigned to the boundary. In the setting of gauge theories, this means that one has to adapt the Batalin–Vilkovisky formalism to a more sophisticated version, called the *BV-BFV formalism*[7] (Cattaneo and Moshayedi 2020).

---

[7] Batalin–Vilkovisky and Batalin–Fradkin–Volkovisky formalism.

Bibliography

Arnold, V.I.: Mathematical Methods of Classical Mechanics. Springer Graduate Texts in Mathematics (1978)

Atiyah, M.F.: Topological quantum field theories. Publ. Math. IHÉS **68**(1), 175–186 (1988)

Batalin, I.A., Vilkovisky, G.A.: Gauge algebra and quantization. Phys. Lett. B **102**(1), 27–31 (1981)

Becchi, C., Rouet, A., Stora, R.: Renormalization of gauge theories. Ann. Phys. **98**(2), 287–321 (1976)

Bogachev, V.I.: Gaussian Measures, vol. 62. Mathematical Surveys and Monographs (1998)

Cameron, R.H., Martin, W.T.: Transformations of wiener integrals under translations. Ann. Math. **45**(2), 386–396 (1944)

Cattaneo, A.S., Moshayedi, N.: Introduction to the BV-BFV formalism. Rev. Math. Phys. **32**, 67 (2020)

Dirac, P.A.M.: The Principles of Quantum Mechanics. Oxford University Press (1930)

Dirac, P.A.M.: The lagrangian in quantum mechanics. Physikalische Zeitschrift der Sowjetunion **3**, 64–74 (1933)

Einstein, A.: Elementare Theorie der Brownschen Bewegung, Zeitschrift für Elektrochemie und angewandte physikalische. Chemie **14**, 235–239 (1908)

Eldredge, N.: Analysis and Probability on Infinite-Dimensional Spaces. arXiv:1607.03591v2 (2016)

Faddeev, L.D., Popov, V.N.: Feynman diagrams for the Yang-Mills fields. Phys. Lett. B **25**(1), 29–30 (1967)

Fernique, X.: Intégrabilité des vecteurs gaussiens. Comptes Rendus de l' Académie des Sciences, Série A-B. **270**, 1698–1699 (1970)

Feynman, R.P.: The Principle of Least Action in Quantum Mechanics, Thesis (Ph.D.). Department of Physics, Princeton University, Princeton, NJ, USA (1942)

Feynman, R.P., Hibbs, A.R.: Quantum Mechanics and Path Integrals. McGraw-Hill, New York (1965)

Glaser, V.: On the equivalence of the Euclidean and Wightman formulation of field theory. Commun. Math. Phys. **37**(4), 257–272 (1974)

Glimm, J., Jaffe, A.: Quantum Physics: A Functional Integral Point of View. Springer, New York (1987)

Groenewold, H.J.: On the principles of elementary quantum mechanics. Physica **12**, 405–460 (1946)

Guichardet, A.: Symmetric Hilbert Spaces and Related Topics. Springer Lecture Notes in Mathematics (1972)

Hall, B.C.: Quantum Theory for Mathematicians. Springer Graduate Texts in Mathematics (2013)

© The Author(s), under exclusive license to Springer Nature Singapore Pte Ltd. 2023   117
N. Moshayedi, *Quantum Field Theory and Functional Integrals*,
SpringerBriefs in Physics, https://doi.org/10.1007/978-981-99-3530-7

Janson, S.: Gaussian Hilbert Spaces. Cambridge University Press (1997)

Johnson, G.W., Lapidus, M.L.: The Feynman Integral and Feynman's Operational Calculus. Oxford University Press (2000)

Kac, M.: On distributions of certain wiener functionals. Trans. Amer. Math. Soc. **65**(1), 1–13 (1949)

Kuo, H.-H.: Gaussian Measures in Banach Spaces. Springer Lecture Notes in Mathematics (1975)

Mörters, P., Peres, Y.: Brownian Motion, vol. 30. Cambridge Series in Statistical and Probabilistic Mathematics (2010)

Osterwalder, K., Schrader, R.: Axioms for euclidean green's functions I/II. Commun. Math. Phys. **31**, 83–112 (1973); **42**, 281–305 (1975)

Prokhorov, Y.V.: Convergence of random processes and limit theorems in probability theory. Theory Probab. Appl. **1**(2), 157–214 (1956)

Reed, M., Simon, B.: Methods of Modern Mathematical Physics I: Functional Analysis. Elsevier (1981)

Schrödinger, E.: An undulatory theory of the mechanics of atoms and molecules. Phys. Rev. **28**(6), 1049–1070 (1926)

Segal, G.B.: The definition of conformal field theory. Diff. Geom. Methods Theor. Phys. **250**, 165–171 (1988)

Simon, B.: Functional Integration and Quantum Physics, vol. 86. Elsevier (1979)

Simon, B.: The $P(\phi)_2$ Euclidean (Quantum) Field Theory. Princeton University Press (1974)

Stone, M.H.: On one-parameter unitary groups in Hilbert Space. Ann. Math. **33**(3), 643–648 (1932)

Takhtajan, L.A.: Quantum Mechanics for Mathematicians, vol. 95. Graduate Studies in Mathematics (2008)

Tyutin, I.V.: Gauge Invariance in Field Theory and Statistical Physics in Operator Formalism, vol. 39. Lebedev Physics Institute preprint (1975)

Weyl, H.: Quantenmechanik und gruppentheorie. Zeitschrift für Physik **46**, 1–46 (1927)

Wick, G.C.: The evaluation of the collision matrix. Phys. Rev. **80**(2), 268–272 (1950)

Wiener, N.: Differential-space. J. Math. Phys. **2**, 131–174 (1923)

Wightman, A.S., Garding, L.: Fields as operator-valued distributions in relativistic quantum theory. Arkiv f. Fysik, Kungl. Svenska Vetenskapsak. **28**, pp. 129–189 (1964)

Zinn Justin, J.: Path Integrals in Quantum Mechanics. Oxford University Press (2004)

Printed in the United States
by Baker & Taylor Publisher Services